The Pocket Book of

BIRD

ANATOMY

First published in 2020

Published by Australian Geographic
Level 7, 54 Park Street, Sydney NSW 2000
Telephone: +61 2 9136 7214

Email: editorial@ausgeo.com.au
www.australiangeographic.com.au

Australian Geographic customer service
1300 555 176
(local call rate within Australia)

Copyright © UniPress Books Ltd 2020
Published by arrangement with UniPress Books Ltd
www.unipressbooks.com

ISBN 978-1-925847-70-3

Printed in China

Australian
GEOGRAPHIC

The Pocket Book of

BIRD
ANATOMY

Marianne Taylor

CONTENTS

INTRODUCTION

In the world today, there are about 10,000 species of birds. They range in size from the 2g (0.07oz) Bee Hummingbird (*Mellisuga helenae*) to the Ostrich, *Struthio camelus*, which weighs in at some 110kg (220lb). There are birds native to every part of every continent, and between them they exploit every kind of habitat from desert and ice cap to the lushest forest. Among them are record-breaking high-fliers, death-defying deep-sea divers, fearsome killers and the most devoted of lovers. Yet despite all of this variety, their anatomical forms show great consistency. Avian traits are instantly and irrefutably apparent in every bird.

As the sole surviving dinosaurs on Earth, birds have a long evolutionary history. Their diversity today is a testament to the success of the avian prototype, which first appeared some 150 million years ago. Feathered dinosaurs existed before this, and feathers were the breakthrough that allowed these animals to maintain their body temperature and thus survive in more inhospitable climates. But it was the adaptation of the forelimb and its feathers into flight-capable wings that allowed the lineage Aves to take off.

Flight opens up a vast array of opportunities that are not available to earthbound animals – most significantly, rapid long-distance travel and the ability to feed and nest in inaccessible places, such as tall treetops, cliff faces and on remote islands. However, flight comes at a cost – in fact, many costs. Birds need a certain power-to-weight ratio to make flight not only possible but energy efficient enough for them to be able to fuel themselves. That means that all bodily structures need to be as light as possible, while also being exceptionally strong to resist the stresses of flight.

The anatomy of a typical flying bird is, therefore, considerably pared down compared to similar-sized mammals and reptiles. Their skeletons contain fewer and smaller (but stronger) bones, their muscles (except for those that drive wingbeats) are slimmed down, and their digestive and waste management processes are extra efficient. Birds live close to their bodily limits and at a fast pace. To cope with this, they have large and complex brains and some show intelligence comparable to that of the

ⓥ The sensory abilities of owls are fine-tuned to help them locate and target their prey in almost complete darkness.

cleverest wild mammals. Their intelligence means that their social interactions are complex, and it is for social reasons that they have evolved the most brilliant colours and most beautiful and varied voices in nature.

The least bird-like of birds are those species that are flightless. All flightless birds have flying ancestors, but they have evolved a way of life where other extreme adaptations offset the loss of flight. Penguins can dive deeper and for longer than any other birds, ostriches are the fastest-running creatures on two legs, and kiwis have a sense of smell that is almost without equal.

This book explores the anatomy of all kinds of birds, from egg to adulthood. We look at the structure and inner workings of all their internal systems from skeleton to skin, respiration to circulation, reproduction to digestion, mind to motion. Through this exploration, we build a picture of the bird as a complete system, a natural machine honed through evolution to survive and thrive and to impress, amaze and enchant.

ⓐ Although they cannot fly, penguins make use of their wings to propel them on longer and deeper underwater dives than any other birds can achieve.

ⓥ Woodpeckers' skulls have exceptional built-in shock absorption, to mitigate the g-force they experience when using their bills to chisel into wood.

1

ANCESTORS AND EVOLUTION

Birds are all that remain today of the dinosaurs. The history of their emergence from the mighty theropod dinosaur lineage is told through millennia of fossils, and evidence of this ancestry is apparent in their living bodies today.

⊙ *Confuciusornis*, a fossil bird from China that lived about 120 million years ago, was crow-sized, with prominent wing claws and a toothless bill.

EVOLUTIONARY TREE

Birds are vertebrates, and they have four limbs. That means birds are tetrapods like humans and other mammals, and also reptiles and amphibians.

The fossil record suggests the first tetrapods date back about 400 million years. Although they resembled any other fishes to the untrained eye, they had lungs as well as gills, and their pectoral and pelvic fins were becoming modified into structures that could support their weight and propel them on land. Earth's land surface was already home to an array of plant life and insects, including the first flying insects, and was ripe for colonisation by larger animals; the world was also recovering from the late Devonian mass extinction (about 360–370 million years ago) which had caused wholesale loss of biodiversity among marine life in particular.

The first tetrapods, even those that could move freely on land, still needed to place their eggs in water. By 312 million years ago, early reptiles had evolved. These were the first amniotes – animals that produced eggs with sturdy shells that could survive and hatch on land. As well as air-proof eggs, they also had adaptations to prevent water loss in their own bodies, such as dry and thicker skin, and more efficient lungs and kidneys.

Some tetrapods have, over time, lost two or all four of their limbs. Snakes, whales and the worm-like amphibious caecilians are examples of tetrapods that no longer have four functional limbs, but they all descend from four-limbed ancestors and their internal anatomy retains traces of this heritage.

THE SAUROPSID LINEAGE

By 300 million years ago, these first reptiles had split into two primary lineages – the sauropsids (the ancestors of dinosaurs, birds and modern reptiles) and the synapsids (the ancestors of mammals). Crocodiles and dinosaurs (including birds) belong to a branch of sauropsids called Archosauromorpha, while modern snakes and lizards descend from another branch, the Lepidosauromorpha. The ancestry of modern shelled reptiles (tortoises and their relatives) is still not fully understood.

Archosauromorpha first appeared some 260 million years ago, and since then has branched and diversified dramatically, though most of those branches were not destined to survive to modern times. About 230 million years ago in the late Triassic period, one of its lineages (Archosaura) split into two pathways – Pseudosuchia and Avemetatarsalia. Species within Avemetatarsalia included the pterosaurs ('pterodactyls') and also many of the familiar dinosaurs, such as *Diplodocus* and other long-necked sauropods, *Triceratops* and its horned relatives. It also included the bipedal theropods – *Tyrannosaurus rex, Velociraptor* and many others; this is the lineage that gave rise to birds. The Cretaceous–Paleogene mass extinction that took place some 66 million years ago killed off most of these animals, though, and modern birds are all that survive today of Avemetatarsalia.

Pseudosuchia is the sister lineage to Avemetatarsalia. It includes a great variety of reptiles, but most of those were wiped out in the earlier Triassic–Jurassic mass extinction, about 200 million years ago. Today, all that remains of Pseudosuchia are crocodiles, alligators and their relatives.

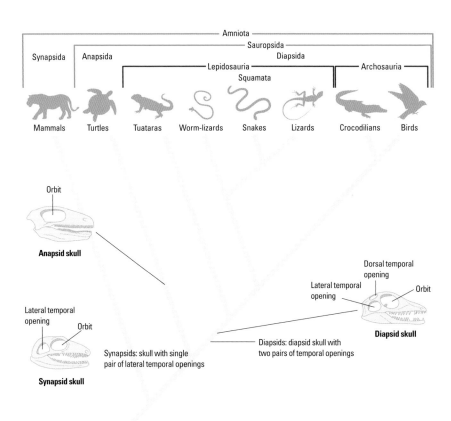

The evolutionary relationships between modern mammals, reptiles and birds.

The earliest members of these two lineages differed most obviously in the bone structure of their ankles (Avemetatarsalia means 'bird ankles') but their surviving modern descendants have followed such different evolutionary pathways and become so different in body shape as a result that it's difficult to believe how closely related they are.

11

THEROPODS

As we have discovered more and more about its anatomy through fossil analysis, the theropod dinosaur of our imagination has evolved from an awkwardly upright, lumbering scaly lizard to something feathered, gracefully swift-moving and altogether more bird-like.

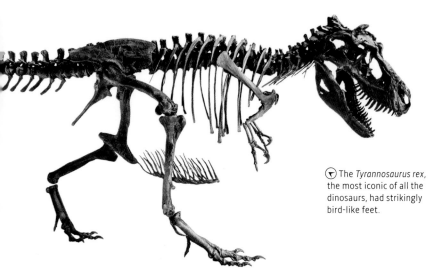

ⵧ The *Tyrannosaurus rex*, the most iconic of all the dinosaurs, had strikingly bird-like feet.

M ost theropods were carnivores, and among them were the biggest land carnivores ever to exist. An adult *Tyrannosaurus rex*, living some 67 million years ago, would have measured about 12–13 metres (39–42 feet) from tail tip to snout, and probably weighed up to 8 tonnes (9 tons). It would have moved in a fairly horizontal stance on powerful hind legs, counterbalanced by a long, thick-based tail, catching and killing prey with its massive jaws, its forelimbs being famously very small for its overall body size. It had light feathering on parts of its body and, despite being a land predator, it had strong but partly hollow bones, a trait common to all theropods.

Other more bird-like theropods included the ornithomimids, which were smaller and more lightly built, with long, partly feathered arms as well as long hind legs. They were small-headed, possibly herbivorous, and would have lived in a similar manner to modern ostriches, running rapidly on large, three-toed feet.

THE MANIRAPTORA LINEAGE

The theropod lineage that includes the modern birds is Maniraptora. Early maniraptorans were small animals with a number of anatomical traits that set them apart from other theropods. They had long arms that ended in just three fingers, and unlike all other dinosaurs they possessed breastbones, rather than sternal plates made of cartilage. Many groups had advanced feather types, including soft down fibres, and also elongated wing and tail feathers, permitting flight. Certain maniraptoran

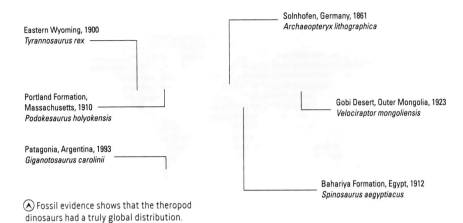

Eastern Wyoming, 1900
Tyrannosaurus rex

Solnhofen, Germany, 1861
Archaeopteryx lithographica

Portland Formation,
Massachusetts, 1910
Podokesaurus holyokensis

Gobi Desert, Outer Mongolia, 1923
Velociraptor mongoliensis

Patagonia, Argentina, 1993
Giganotosaurus carolinii

Bahariya Formation, Egypt, 1912
Spinosaurus aegyptiacus

(A) Fossil evidence shows that the theropod dinosaurs had a truly global distribution.

dinosaurs were probably capable of true flight – among them the genus *Eosinopteryx*, which had long, strong-shafted wing and tail feathers. There were also four-winged forms such as *Microraptor* and *Changyuraptor*, with flight feathers on fore and hind limbs.

Some 160 million years ago, the non-avian maniraptoran family Scansoriopterygidae had elongated and strongly clawed third fingers, which supported a fleshy patagium, or membrane, like that which forms the wings of bats. This suggests that they probably lived in trees, climbing and gliding between them.

Velociraptor was an avian maniraptor that lived some 75 million years ago. Its size, structure and projected weight suggested it pursued a terrestrial lifestyle. However, it had quite long-feathered forelimbs, which it may have flapped to provide extra power when running hard. It may also have used its wings for stability while pinning down prey.

The theropods form a highly diverse group of dinosaurs, which first came into being some 230 million years ago and survives to the modern day in the form of birds. We have assigned more than 1,100 theropod fossils to distinct species, but today we recognise more than 10,000 species of birds. Given the great rarity of fossils generally, it is likely that the total number of theropod species that have existed over the group's long lifespan is many times larger than these two sums combined.

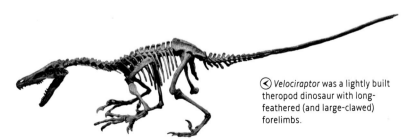

(<) *Velociraptor* was a lightly built theropod dinosaur with long-feathered (and large-clawed) forelimbs.

13

EARLY BIRDS

There is no simple set of criteria that allows us to definitively classify a particular maniraptor as a 'true' bird. The process of transition from more classically reptile-like body types to more bird-like ones is gradual, although in evolutionary terms it was also relatively rapid at times.

We cannot be certain that any particular prehistoric theropod, however bird-like, was definitely an ancestral species of modern birds; however, we can see the emergence of modern avian traits through study of their fossils. The best-known fossil 'bird' is *Archaeopteryx*, which lived some 150 million years ago. Several well-preserved fossils have been found in what is now southern Germany. It was fully feathered, with well-developed wing and tail feathers, and was probably capable of some degree of flight; its superbly preserved fossils reveal an unmistakably bird-like outline. However, its long tail was more reptile than bird, being formed by about 20 separate caudal (tail) vertebrae, and the tail feathers grew out of it along its length, like the fronds of a fern. Modern birds lack a bony tail and their caudal vertebrae are fused into a small triangular plate called the pygostyle, which supports a fan of tail feathers. Other non-avian traits possessed by *Archaeopteryx* included fully developed sharp teeth, no ridge or keel on its sternum, and three projecting, clawed fingers at the bend of the wing.

At the time of its discovery in the 19th century, nothing like *Archaeopteryx* had ever been found before. Several of its more bird-like traits were not thought to exist in dinosaurs at all, but have since been discovered in several theropods and even in some non-theropod dinosaurs. These include fully developed feathers, the existence of a furcula or wishbone (formed by fused clavicles), and a partly reversed first toe. Nevertheless, *Archaeopteryx* is usually considered to be a true (if primitive) bird rather than a bird-like dinosaur.

The eleven known specimens of *Archaeopteryx* are similar enough that most paleontologists assign them all to just one species – *Archaeopteryx lithographica*. The second word of its name means 'written in rock'.

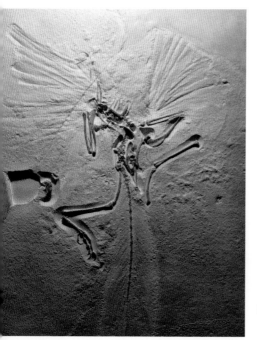

⊲ The well-preserved forelimb feathers of this *Archaeopteryx* fossil leave no doubt that this creature had flight-capable wings.

SIGNS OF THE MODERN BIRD

Fossils from the Cretaceous period (140–66 million years ago) show early birds diversifying into a variety of forms, some of them flying and others flightless. One significant group was the Enantiornithes, or 'opposite-birds', whose shoulder joints articulate the other way round to modern birds, leading to a different manner of flight. These birds also had teeth and clawed wings, but had evolved a pygostyle and reversed hind toe, and in many ways resembled modern bird toes.

Further changes towards modern bird anatomy took place, including the change from bony tail to pygostyle, the reduction of forelimb digits and the loss of teeth. Other late Cretaceous birds include the gull-like

(∧) This *Ichthyornis* skeleton shows its overall similarity to modern birds, but note the backwards-angled teeth, which allowed it to grip its fish prey.

Ichthyornis, which had a bird-like bill but retained a few sharp teeth in the middle of each jaw, and the huge diver-like Hesperornithes, a foot-propelled, flightless fish-catcher with small, sharp teeth and vestigial wings. All of these forms disappeared in the Cretaceous–Paleogene mass extinction event, but some of their relatives survived and gave rise to modern birds through the Paleogene and Neogene eras.

LOST ODDITIES

The dinosaurs and many other land and sea animals were wiped out in the Cretaceous–Paleogene mass extinction, some 66 million years ago. In the following millennia, the surviving birds and mammals spread and diversified rapidly, occupying ecological roles once performed by dinosaurs.

Some of the lineages that evolved over this time have survived to the present day, although individual species rarely persisted for longer than a million years. Those birds that have come and gone include many remarkable species that coexisted with humankind. One of the earliest taxa of birds to appear in the fossil record is Palaeognathae – the ratites and tinamous. While tinamous are small, weakly flying birds that resemble chickens, the ratites are mostly very large, flightless birds, including the largest living bird species, the Ostrich. They are distinguished by the lack of a sternum keel, among other traits.

The first ratites existed at least 56 million years ago, probably originating in Eurasia and from there spreading to Australasia, Africa and the New World. Extinct ratites include the elephant birds of Madagascar (family Aepyornithidae), which weighed in

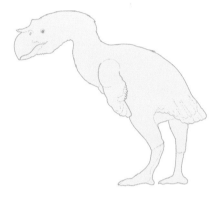

⌃ A real heavyweight among the early birds, *Dromornis* occupied a similar ecological role to that of mammals such as rhinos and elephants.

⌄ Relative sizes of some extinct flightless birds. From left: Giant Moa, Elephant Bird, *Dromornis*, Colossus Penguin and Moa-nalo.

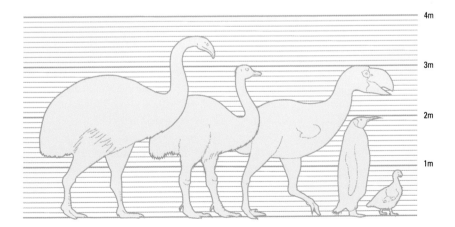

4m

3m

2m

1m

excess of 500kg (1,100lb). The nine species of moa in New Zealand included *Dinornis*, the giant moas, which stood more than 3.5 metres (11 feet) tall, making them the tallest birds ever to have lived. Smaller moas were hunted by the 15kg (33lb) Haast's Eagle (*Hieraaetus moorei*), the world's heaviest ever eagle. All of these species became extinct in the 12th and 13th centuries, not long after humans colonised their respective habitats.

ANCESTORS

Of similar body size to the elephant birds was *Dromornis*, the 'demon duck' of Australia, which looked ratite-like but was a relative of the ancestors of modern ducks and geese. Like the ratites, it ran on two strong legs, and was herbivorous – the bird equivalent of large grazing or browsing mammals. The moa-nalo, heavy flightless birds with large, serrated bills, were also related to modern ducks, and were the largest herbivores on the Hawaiian Islands. *Dromornis* last lived some 30,000 years ago, but the moa-nalo survived until humans colonised Hawaii, sometime after 124 CE.

The penguins are another ancient lineage of birds, first appearing about 62 million years ago, and diversifying greatly over the next 30 million years. Among the more remarkable species was *Palaeeudyptes klekowskii*, the Colossus Penguin. This bird had a long spear for a bill and stood at least 1.6 metres (5 feet) tall; it probably weighed about 115kg (250lb).

With a 6.4-metre (20-foot) wingspan and a bill full of pointed pseudo teeth, *Pelagornis sandersi* was an impressive predatory seabird that lived some 25 million years ago in what's now North America. The only other bird

with a wingspan to rival it was *Argentavis magnificens*, a large scavenging bird present in what is now Argentina, about 7 million years ago. The former species resembled modern albatrosses but was most closely related to ducks and geese. The latter was a relative of modern American vultures.

Today, only a few large, flightless birds survive, but they include successful species such as the Ostrich of Africa and the two species of rhea that occur across much of South America.

Ⓛ The giant moas were nearly twice as tall as an average human being.

CONVERGENT EVOLUTION

In our first attempts to classify the natural world, we grouped bats with birds, and whales with fish. While their similarities are obvious, some of them are also superficial.

As we have learned more about the processes that cause evolution, we have come to understand that unrelated animal groups with similar ways of life have often developed some anatomical similarities too, via different pathways. Examples of convergent evolution, as this phenomenon is known, exist between different bird families and also between birds and other classes of animals. True, powered flight has evolved independently in three modern groups of animals – insects, birds and bats. A fourth group, the extinct pterosaurs, were the first vertebrates to evolve flight, and this lineage also produced the largest ever flying animals. All four groups possess moveable wings to create lift, and have much less body weight than their land-bound equivalents.

In the three vertebrate groups that fly, the forelimbs are modified into wings, with a large, strong and air-resistant but very lightweight surface area to capture and push against air. This surface is a skin membrane (patagium) running from wing tip to toe tip in the case of bats and pterosaurs, and a layered arrangement of feathers in birds. In birds and pterosaurs, all digits but one are reduced in size compared to most other tetrapods, with the single elongated finger forming part of the leading edge of the wing, while in bats the fingers are all greatly elongated and splayed, providing additional support to the patagium. The bird respiratory system includes air sacs and air spaces in their pneumatised bones, and the same was true of pterosaurs, but bats lack these adaptations.

Where birds have evolved in the absence of mammals, they have often diversified to exploit niches that are the preserve of

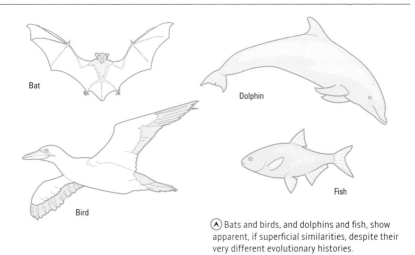

Bat

Dolphin

Bird

Fish

⌃ Bats and birds, and dolphins and fish, show apparent, if superficial similarities, despite their very different evolutionary histories.

(A) Swifts and swallows have remarkably similar body shapes and lifestyles, despite their disparate evolutionary history.

(A) Penguins and puffins (*Fratercula arctica*) are not related, but both have evolved to use their wings for propulsion underwater. However, only auks have retained the ability to fly.

mammals elsewhere. The kiwis of New Zealand – flightless, with stout, robust bodies and a finely tuned sense of smell via nostrils placed near the bill tip – are ecologically similar to hedgehogs and other terrestrial, nocturnal mammals that sniff out their prey in the woodland understorey.

SHARED TRAITS

Convergent evolution across separate bird families is commonplace. Well-known examples include the swifts and swallows, both of which specialise in hunting flying insects. They are fast fliers with streamlined bodies and have small bills with wide, bristle-lined gapes. Swifts are related to the hummingbirds, while swallows are songbirds, more closely allied to larks and warblers. Among seabirds, the penguins and auks are unrelated but show remarkable convergence as deep-diving, wing-propelled fish hunters, clumsy on land but masterly on and in the water. Unlike the penguins, the auks have retained the power of flight, except for the most penguin-like auk of all, the Great Auk (*Pinguinis impennis*), which was hunted to extinction in the 19th century.

ISLAND ENDEMISM

Natural selection shapes a population of living things to best function in their environment and exploit particular ecological niches. In environments where there is a very limited fauna, such as small, remote islands, we find some extreme examples of adaptation.

Flight is a near-universal bird trait, allowing birds to escape many predators and to travel easily to where they need to be. However, being able to fly imposes its own limitations. Flying birds must keep a low body weight and have certain body proportions. Having large wings makes it less easy to move on the ground. Being small-bodied with very high energy needs can make it difficult to retain body heat. Growing a full set of large flight feathers each year is costly on bodily resources. These and other challenges go hand in hand with maintaining a flight-capable body. So a bird species living on a small island and nowhere else (an island endemic) may be pushed along the path to flightlessness if it can make its living from the resources that island holds, and if it doesn't need to escape from predators.

(⌃) The flightless Takahe of New Zealand is the largest and heaviest member of the family Rallidae.

Flightlessness in Rallidae

In the land-bird family Rallidae (rails and crakes), the only sizable land bird family with many flightless members, there are about 142 species. Of those, 20 are flightless, and 16 of those 20 (80 per cent) flightless species are island endemics (living on small or medium-sized, historically predator-free islands). Of the 122 flying species, only 5 per cent are island endemics.

Since 1500 about 30 species of Rallidae (nearly all of them flightless island endemics) have become extinct.

Flightlessness in Rallidae

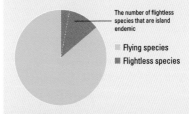

The number of flightless species that are island endemic

- Flying species
- Flightless species

Species of Rallidae that can fly

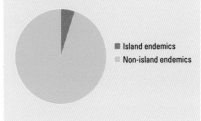

- Island endemics
- Non-island endemics

So there is a tendency for birds endemic to predator-free islands to become poorer fliers, better walkers, and also to become larger in body size (island gigantism). This makes them more energy-efficient, needing fewer calories to survive and function. Examples include New Zealand species like the Takahe (*Porphyrio hochstetteri*; a rail) and the Kakapo (*Strigops habroptilus*; a parrot) – both of these flightless birds are exceptionally big and heavy examples of their respective families.

Lesser-known island endemics include the Pitcairn Reed Warbler (*Acrocephalus vaughani*), one of several reed warbler species endemic to particular South Pacific islands.

Ⓐ The flightless Kakapo of New Zealand is the world's heaviest parrot species, reaching up to 4kg (8³/₄lb) in weight.

Ⓐ Most known members of the swamphen genus *Porphyrio* are extinct or endangered. The Purple Swamphen (*Porphyrio porphyrio*) remains relatively widespread in the Old World.

Many of these warblers have limited flight capabilities, because of a lack of predators, and are bigger than their more widespread cousins. The Pitcairn Reed Warbler, the only native songbird on its island, also has a very high incidence of partial leucism (random patches of white feathering, caused by a genetic mutation) – a conspicuous trait that tends to be 'weeded out' quickly in populations of birds that do have natural predators.

Unanticipated predators

Island endemics, while well-adapted to their environment, are extremely vulnerable to chance events, such as extreme weather and the arrival of non-native predators. The Dodo (*Raphus cucullatus*), native to Mauritius, was a large, flightless pigeon that was rapidly wiped out when people (along with dogs, pigs and rats) arrived on the island. Many other island endemics became extinct in the same way, and nearly all others are in need of drastic conservation measures if they are to survive.

EVOLUTION TODAY

By its nature, evolution is unending. The process of natural selection shapes the anatomy, physiology and behaviour of a population of animals to thrive better in that environment, by weeding out individuals least able to survive, leaving the 'most fit' alive to pass on their genes to a new generation.

As environments change, so successive generations adapt (as long as environmental change is not too rapid – if it is, then extinction is a more likely outcome than adaptation). The process is usually extremely gradual, but sometimes much more rapid and even at times observable within our own lifespans. When new opportunities are suddenly available to a species, we may observe a phenomenon called adaptive radiation. This has famously occurred on the Galápagos Islands. These volcanic islands are geologically young and have only really been hospitable to animal life for about 1.6 million years. The few land birds they support include some 15 species of finch-like tanagers, all similar but adapted to slightly different habitats and lifestyles. They differ particularly in bill shape – some have thick bills for seed cracking, while others have slimmer bills, suited to an insectivorous diet. Studying the Galápagos finches helped Charles Darwin formulate his theory of evolution – he concluded that all species descended from a recent common ancestor that reached the islands, and underwent relatively rapid adaptive radiation, with different lineages branching out to exploit the range of ecological niches available.

Ⓐ Garden bird-feeding stations in the UK are helping to drive evolutionary change in the Blackcap.

THE EVOLVING BLACKCAP

When two parts of a species' population begin to differentiate, this may sometimes be the start of speciation – one species becoming two (or more). This process is occurring in the Blackcap (*Sylvia atricapilla*),

a warbler that breeds in northern and eastern Europe and migrates to Iberia and northern Africa in winter. In the 1960s, birdwatchers in the UK noticed that growing numbers of blackcaps were present in winter. Ringing studies (marking wild birds with unique leg rings, to allow future identification of individuals) revealed that the wintering birds had travelled from eastern Europe. A single genetic mutation had changed the direction the birds take when migrating from south-west to north-west, sending them to

Ⓐ The Large Ground Finch (*Geospiza magnirostris*), a seed-eating species, is the biggest-billed of all of the Galápagos finches.

Ⓐ The pointed, strong bill of the Common Cactus Finch (*Geospiza scandens*) is adapted to take seeds and fruit from prickly-pear cacti, a common Galápagos plant.

Ⓐ The Woodpecker Finch (*Camarhynchus pallidus*) is probably the most famous Galápagos finch. It probes tree holes for insects and will also use a cactus spine as a tool to extract prey.

Ⓐ The insect-eating Green Warbler Finch (*Certhidea olivacea*) has a smaller and slimmer bill than the other Galápagos finches.

the UK. The new trait spread quickly through the population, as the UK proved a very suitable place for the birds to overwinter (thanks, in part, to so many people putting out birdfood for garden birds in winter).

Now, the Blackcap population in eastern Europe is split into two distinct groups based on where they go in winter, and the UK-wintering birds are beginning to show different anatomical traits as they become better adapted to their new winter home and somewhat different winter diet. They

have shorter wings and browner plumage, and slimmer bills for a more varied diet – those that winter in Africa and Iberia have stouter bills and eat mainly fruits in winter. There is also a strong tendency for UK-winterers to breed with other UK-winterers, as they return from migration slightly earlier than the Africa-winterers, so the two forms are starting to experience reproductive isolation – a strong driver of evolutionary change.

MUTATION AND NATURAL SELECTION

The processes of genetic mutation and natural selection are what drive evolutionary change. This is how populations of living things adapt to their environment, and over time this can involve changes to their anatomy and physiology – and sometimes also their behaviour.

Bodies are built according to the instructions in the genes. This means that genetic changes (mutations) produce variety in bodies and behaviour, but it is natural selection that 'decides' which individuals prosper and which fail. Natural selection would have come into play as soon as self-replicating life began on Earth. Not all organisms are equal, and those that function less well are less likely to replicate. Early life reproduced just by one cell dividing into two. Later, sexual reproduction appeared, with two organisms combining their genes to produce offspring that carry a mixture of both parents' genes and therefore their traits. With more variety introduced into the population, the speed of natural selection would have increased.

(A) Genetic mutations are responsible for the array of colour forms that you may observe in a typical flock of feral pigeons.

Cell division

Every cell in an organism's body carries the same set of paired chromosomes, each of which carries hundreds or thousands of genes, all made from the molecule DNA. Before a cell goes through the process of dividing into two (mitosis), it makes new copies of all of its chromosomes, so that when it divides, the two 'daughter cells' have exactly the same DNA as the original cell – but in practice, small errors in the replication process often occur.

The precursors to egg and sperm cells go through two cell divisions before they reach their final forms, and this process (meiosis) is slightly different. In the first division, some portions of a chromosome swap sides, moving from one of the pair to the other. Then, in the second division, only one half of each chromosome pair is passed on to the two daughter cells. This means that when an egg and a sperm cell meet, they each provide 50 per cent of the DNA for each chromosome. The resultant embryo has a random mix of genes from each parent, along with any new mutations from errors introduced during the process.

Survival of the fittest

So, no two individuals in a population of a sexually reproducing organism have the same genes, and overall the population has a lot of genetic variety. Not all individuals can survive and breed, only those best adapted to their environment, and if environmental conditions are changing, the best traits for survival will also change. Sometimes in a population, two different approaches to life are both successful and natural selection will drive a split – from one species into two. Evolution happens most quickly when new opportunities become available, and when new challenges to survival are introduced.

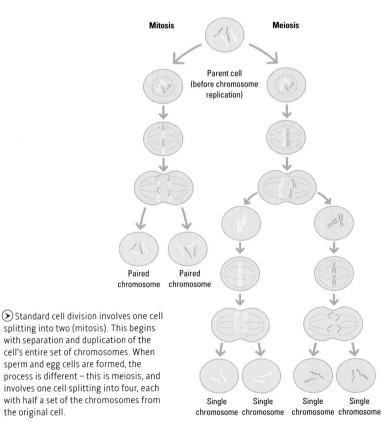

Mitosis

Meiosis

Parent cell
(before chromosome
replication)

Paired
chromosome

Paired
chromosome

Single
chromosome

Single
chromosome

Single
chromosome

Single
chromosome

(>) Standard cell division involves one cell splitting into two (mitosis). This begins with separation and duplication of the cell's entire set of chromosomes. When sperm and egg cells are formed, the process is different – this is meiosis, and involves one cell splitting into four, each with half a set of the chromosomes from the original cell.

CELL BIOLOGY

Cells are the living, functional units of an animal's body and understanding the fundaments of cell biology is helpful for understanding anatomy as a whole.

Each cell is a miniature machine that carries out varied tasks, depending on their type and position in the body. While they communicate with other cells, they are, in a certain sense at least, self-regulating, and some move and operate almost entirely independently of other cells. Most are microscopic. The largest cell in the human body is the ovum or egg cell, at 0.1mm (0.0004in) across, while technically the yolk in a bird's egg is one single, giant cell.

A typical animal cell is a bag of fluid (cytoplasm), enclosed within a cell membrane that is semipermeable – it allows certain molecules to pass through it, either freely or at particular points under certain conditions. Incoming molecules are those needed for the cell to function, while the molecules it

secretes are manufactured in the cell and travel elsewhere in the body (most often in the bloodstream) to carry out other functions. The cytoplasm contains the cell's organelles – smaller structures with specific functions.

Under a microscope, the most obvious organelle is the nucleus – it looks like a round, dark spot. Some large cells have multiple nuclei. This part of the cell stores its genes, in the form of a molecule called DNA. Each gene is the 'recipe' for making a particular kind of protein, and each full strand of DNA, containing thousands of genes, is a chromosome. Every cell nucleus contains a full set of chromosome pairs, and these are identical in every cell of the body. The number varies by species but in most birds there are 38 or 40 pairs of chromosomes.

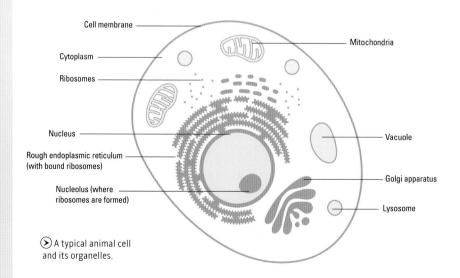

Cell membrane
Cytoplasm
Ribosomes
Nucleus
Rough endoplasmic reticulum (with bound ribosomes)
Nucleolus (where ribosomes are formed)
Mitochondria
Vacuole
Golgi apparatus
Lysosome

⊙ A typical animal cell and its organelles.

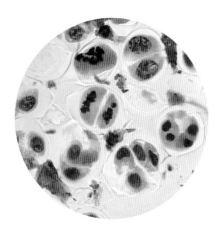

⊘ Animal cells undergoing division (mitosis) viewed under a microscope.

including regulating the whole cell's metabolism. They have their own membranes and their own DNA, which they use to build the proteins they need for the processes they carry out.

Other structures within a cell include the centrosomes, which assist with cell division, the Golgi apparatus, which packs newly formed proteins into membranes before they are secreted by the cell, and lysosomes, which break down waste products within the cytoplasm.

Shape variations

Though most cells are round in shape, there is some variation depending on location and function. The sperm cell has a long whip-like flagellum or tail to enable it to swim, with numerous mitochondria at the tail's base to provide the necessary energy. Neurons or nerve cells have a long, narrow extension (axon) to transmit the nervous impulse. Skeletal muscle cells are long and narrow, containing organised chains of proteins that can break and reform their bonds in different positions to allow muscular contraction. Phagocytic white blood cells are blob-shaped but soft and flexible – able to dramatically distort their membranes to engulf other cells.

When a cell divides, producing two copies of itself, the process begins in the nucleus, with the chromosomes being duplicated first, before being pulled apart into separate parts of the nucleus. The nucleus then splits, retaining a full set of chromosomes in each half, and the rest of the cell divides around each new nucleus, resulting in two complete and identical new cells. The exception is when new sperm and egg cells are produced – these only have half a chromosome set each.

Ribosomes are small organelles that may be freely moving in the cytoplasm, or bound to a membrane called the endoplasmic reticulum, which is continuous with the membrane of the nucleus. The function of ribosomes is to build proteins, according to instructions from the genes in the cell nucleus.

Mitochondria are sausage-shaped organelles that release energy from glucose (and some other nutrients), through aerobic respiration, and have certain other functions,

2

THE SKELETON

One of the most vital adaptations to life on the wing that has evolved in modern birds is a skeleton that is strikingly pared-down but still extremely strong.

⊘ The bird skeleton is all neck and legs, with a unique keeled breastbone on which its flight muscles are anchored.

THE BIRD SKELETON

The lives of most flying birds are fast-paced and make tremendous physical demands upon them. A robust framework is essential, but the bird skeleton also needs flexibility and to be light enough for flight.

Solid bone weighs about 1.6 grams per cubic centimetre, making it much more dense than muscle (1.06 grams per cm³), fat (0.9 grams per cm³), or blood (1.04 grams per cm³). Although solid bird bone is actually more dense than that of equivalent-sized mammals, many weight-saving adaptations are apparent in the bird skeleton.

The most striking thing about a bird skeleton is probably the length of the neck and legs, even in birds that we think of as relatively short-necked and short-legged, such as owls. This illustrates just how much a bird's feathers disguise its true body outline, especially when the bird is resting.

Other obvious differences between the bird skeleton and that of other tetrapod animals is the lack of a chain of tail vertebrae

and the lack of teeth. The tail vertebrae are reduced to the fused and greatly shortened pygostyle. Tooth enamel is very dense (about 2.7 grams per cm³) so dispensing with it makes for a considerable weight saving. Although the bill can be very large, the bone that forms it is thin – it gains additional rigidity from its lightweight sheath of keratin (the same material from which our fingernails are made). The keel on the sternum is another distinctive bird trait – this long, flattened ridge of bone provides an attachment point for the powerful pectoral muscles that power flight (and swimming, in the case of wing-propelled diving birds such as penguins). Bird vertebrae are also more compact than those of other vertebrates.

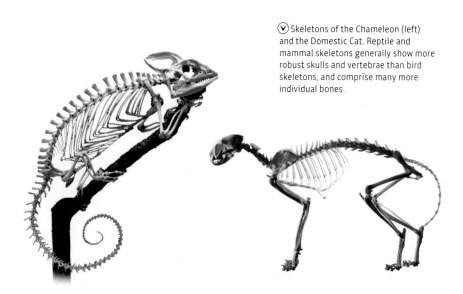

ⓥ Skeletons of the Chameleon (left) and the Domestic Cat. Reptile and mammal skeletons generally show more robust skulls and vertebrae than bird skeletons, and comprise many more individual bones.

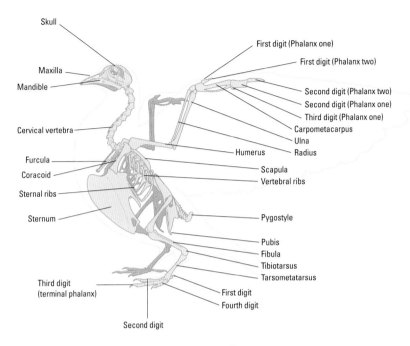

Skull

First digit (Phalanx one)

First digit (Phalanx two)

Maxilla

Second digit (Phalanx two)

Mandible

Second digit (Phalanx one)

Third digit (Phalanx one)

Carpometacarpus

Cervical vertebra

Ulna

Humerus

Radius

Furcula

Scapula

Coracoid

Vertebral ribs

Sternal ribs

Sternum

Pygostyle

Pubis

Fibula

Tibiotarsus

Tarsometatarsus

Third digit
(terminal phalanx)

First digit

Fourth digit

Second digit

(A) The principal bones and bone groups
of a typical flying bird.

SKELETAL ECONOMY

Fusion of some bones and loss of others is another feature of the avian skeleton. The exact number of bones varies greatly between species – the number of neck vertebrae is particularly variable, ranging from 11 to 25. However, birds have fewer bones in total than mammals (for example, a chicken has 120, while a cat has about 230). The collarbones are fused, forming a single bone (the furcula or wishbone). Many bones in the lower foot have fused together to form the single tarsometatarsus. The digits in the wing are reduced (by fusion of some bones and loss of others) to just three, of which only the second digit is of any appreciable size. Some of the lower back vertebrae are fused into a bone called the synsacrum, and the pelvic girdle is also fused to this bone.

Finally, many birds' bones are pneumatised (contain air spaces). This reduces their weight and also connects the bones to the bird body's complex air exchange system.

BIRD DIVERSITY AROUND THE WORLD

Birds live everywhere on Earth, from pole to pole and on every landmass. And even though all birds need land to breed, some seabirds will wander the most remote ocean regions when they are not nesting.

However, some parts of the world and some kinds of habitats are far richer in birdlife than others, in terms of overall numbers, variety of species, and variety of body types and ways of life. There are also dramatic changes in birds' distributions from season to season. The richest habitats on Earth are those found in tropical regions, particularly in central South America but also in eastern Africa and South-east Asia. Of the five countries with the greatest number of bird species, four are in South America (Colombia, Brazil, Peru and Ecuador). The fifth is Indonesia, in South-east Asia. Its species count is greatly boosted by its 18,000 or so islands, many of which have endemic species, found nowhere else on Earth. These top five countries all support more than 1,500 bird species, while Bolivia, Venezuela, China, India, the Democratic Republic of the Congo, Mexico, Tanzania, Kenya and Argentina all hold more than 1,000 each.

Centre of diversity

When you look at the distribution of a bird family around the world, it is often apparent where that family originated by their 'centre of diversity' – the area where they are represented by the widest range of species.

Bird species diversity worldwide

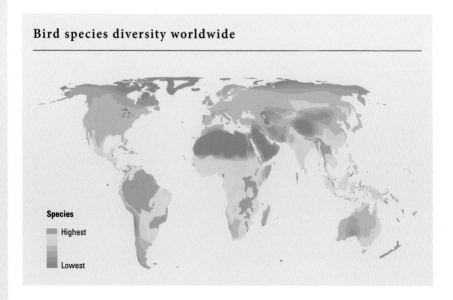

Species

Highest

Lowest

ⓒ Although very few people have ever seen one, Wilson's Storm Petrel is one of the world's most numerous bird species.

ⓥ The forests of South America are home to about 40 per cent of the world's parrot species, including the dazzling Scarlet Macaw (*Ara macao*).

Usually, diversity gradually diminishes the farther you travel from this centre. The wren family, for example, includes about 88 species and is most diverse in Central America and north-western South America. There are 35 species in Colombia in a great variety of shapes and sizes. However, there are only 11 in the United States and nine in Canada. Diversity also decreases as you head south, with just four wren species in Argentina. Only one species is found in the Old World, but it is present across a huge area, from western Europe east to China and south to North Africa.

Temperate areas support fewer birds, and polar regions the fewest of all. Only 45 species of birds have been recorded on the Antarctic continent, and 246 on Greenland (the vast majority as 'accidentals', rather than occurring regularly). However, numbers of individual birds can still be very high. For example, Wilson's Storm Petrel (*Oceanites oceanicus*), which breeds on the Antarctic coast and nearby islands, is one of the world's most abundant species, with up to 10 million pairs.

SKULL

The skull provides protection and support for the brain, eyes, inner ears and various other delicate structures. A bird skull is recognisable as such by the presence of the bill and absence of teeth, and the very large orbits, or eye sockets.

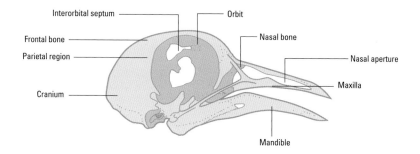

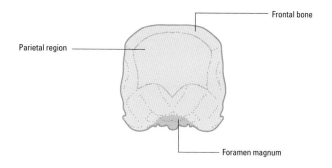

A skull is formed by several overlapping smaller bones, and also includes the separate lower jawbone and sclerotic eye-rings (rings of small bones around each eye). Viewed from the side, the spaces in the skull for the bird's eyes can be seen to almost meet in the middle, separated only by a thin plate of bone (the interorbital septum). The braincase sits behind the orbits and takes up, in most species, considerably less space than they do – in the Ostrich, each eyeball is famously a little larger than its brain. A hole at the base of the braincase, the foramen magnum, is where the spinal cord exits the head, along with veins and arteries.

THE BILL

In mammals, the nostrils mark the tip of the snout, and the nose often extends beyond the jaw. But in birds the tips of the jaws are elongated to form the bill, which extends a long way forwards of the nostrils. The top part of the bill, which is continuous with the rest of the skull, is the maxilla, and the lower part of the bill is the outermost part of the mandible – the lower jawbone. The maxilla is

(>) The prominent sclerotic rings that surround the eye sockets are formed of tiny, delicate bone plates.

(<) The principal bones and bone regions of a typical bird skull.

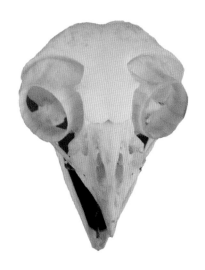

mostly hollow with a large hole on each side, corresponding to the nostril. The mandible is V-shaped, the two forks articulating with the back of the skull.

The frontal bone forms the top of the skull, and the nasal bone the part just in front of the orbits. Because these two bones connect with a hinge rather than being fused, a bird can move its upper jaw independently of the rest of its skull. This movement, known as prokinesis, enables the bill to open very wide (you can see it happening when a bird yawns). Some long-billed birds, such as curlews and sandpipers, can also flex the upper bill tip upwards – this is rhynchokinesis, and is thought to help them catch and manipulate small prey items in the bill more efficiently.

THE SCLEROTIC EYE-RING

The ring of tiny, overlapping bone plates that surrounds the eye is present in many birds. It is particularly large and prominent in owls and some other birds of prey, and in herons. However, it is found in smaller birds too, even hummingbirds, in which the individual bones are no thicker than tissue paper. The ring is thought to provide some support to the eyeball – albeit flexible support to allow for changes in eyeball shape as the eye's focal point changes from close-up to distant objects. The ring is more rigid in birds that dive deeply underwater, where the eye will be subjected to considerable water pressure.

(v) A cormorant demonstrates the flexibility of its upper bill tip.

WING BONES

A bird's wing corresponds to a quadrupedal mammal's foreleg or the human arm. In most species, the wing's main bones are longer, heavier and denser than their equivalents in the hind limb, reflecting their roles as the main supporting structures of limbs used for powered flight.

The obvious bend of a bird's wing – the carpal joint – corresponds to the human wrist, while the elbow joint is located much closer to the body. Its position is hard to spot in a living and feathered bird. The bone above the elbow, connecting to the shoulder, is the humerus, while between the elbow and the wrist are the ulna and radius. These two long bones meet at either end but have a sometimes quite large elliptical space between them, which changes shape as the bones bend during the wingbeat cycle. The ulna is usually stouter than the radius (in mammals, it is the other way around).

THE BIRD HAND

The wrist and 'hand' of a bird contain far fewer bones than those of mammals and reptiles. Through the evolutionary process that transformed a foreleg into a wing, some of the original bones have been lost, and others fused together. The carpometacarpus looks like two long bones that are fused at both ends but not in the centre. It is analogous to the carpal (wrist) and metacarpal (body of the hand) bones in a human hand. The human wrist contains eight carpal bones and the hand has five metacarpals, but in birds just a couple of tiny carpal bones remain besides the carpometacarpals; these are the radiale and the ulnare, involved in the folding of the wrist joint.

Birds have just three digits in the wing, having lost the first and fifth from the original tetrapod forefoot. The first of their digits forms the alula, located on the wrist joint at the leading edge of the wing, and bears a small fan of stiff feathers, like a miniature accessory wing. The alula contains one or two small phalanges (finger bones). It is often not obvious, but it is pushed forwards and its feathers fanned out for certain kinds of flight manoeuvres.

(v) Falcons and other agile fliers make much use of their alulae when carrying out fast brakes and turns.

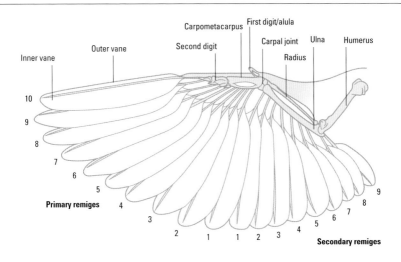

Inner vane
Outer vane
Carpometacarpus
First digit/alula
Second digit
Carpal joint
Ulna
Humerus
Radius
10
9
8
7
6
5
4
3
2
1
Primary remiges
1
2
3
4
5
6
7
8
9
Secondary remiges

The two phalanges of the middle digit connect to the carpometacarpus. Together, these three bones are relatively long and make a bony support for the long primary flight feathers, or remiges, that form the outer part or 'hand' of the wing. The innermost digit consists of just one very small phalanx, which sits close to the second digit's innermost phalanx.

The larger wing bones have strong, dense bone material in their outer walls, but have air spaces inside, which reduces their weight and also connects them to the air sac system in the bird's chest. The bones of a bird's wing are smaller and shorter than you might expect, given the amount of work a wing has to do. The proportions of the bones in a bat's wing could hardly be more different, and show how evolution has come up with two very disparate ways to achieve the same result.

(∧) Avian wing bones are relatively small but provide support to the fan of primary and secondary flight feathers (remiges).

(∨) When a pale-feathered bird such as a gull is backlit in flight, the radius and ulna bones in its wings are often clearly visible.

LEG AND FOOT BONES

Birds use their hind limbs for a variety of purposes. Depending on species, the legs and feet may be adapted to perch, walk, run, hop, swim, climb, cling, dangle or leap.

Hind limbs may also be used to seize or stamp on prey and manipulate food items, and to help preen the plumage, carry items from place to place, fight with rivals, and dig out a nesting scrape or burrow. This variation is reflected in the foot and leg anatomy, including the bones they contain. The standard tetrapod hind limb consists of a femur (thighbone) connected, via the knee joint, to a pair of long bones, the tibia and

⤵ The bend in a flamingo's leg looks wrong until we realise that we are looking at an ankle rather than a knee.

fibula, which sit together much like the radius and ulna in the forelimb. In an intact, feathered bird, the thigh and the knee joint are not usually apparent. The bend we see in the visible part of a bird's leg, usually near the top of it, is the ankle joint (which is why birds' legs appear to bend the 'wrong' way). The part of the leg between the knee and the ankle is also mostly hidden under the plumage in all but very long-legged birds. This part of the leg contains the tibiotarsus, a bone formed by a fusion of the tibia, the fibula, and some foot bones.

The main visible part of the leg is analogous to the body of our foot, which contains our five long metatarsal bones, one corresponding to each toe. In birds, this part contains just one long bone, the tarsometatarsus, formed by the fusion of metatarsals and some other foot bones.

BIRD TOES

Birds usually have four toes, the fifth always being absent. Several species from various families have three rather than four toes, most having lost digit 1 but in a few cases digit 4. The Ostrich has just two (digits 3 and 4). In most species, the first digit (analogous to our big toe) is rotated to point backwards and so is usually referred to as the hind toe. The second, third, and fourth toes (counting from the inside outwards) point forwards. The toes contain small bones, phalanges, which vary in number but typically there are two in the hind toe, three in the second toe, four in the third and five in the fifth.

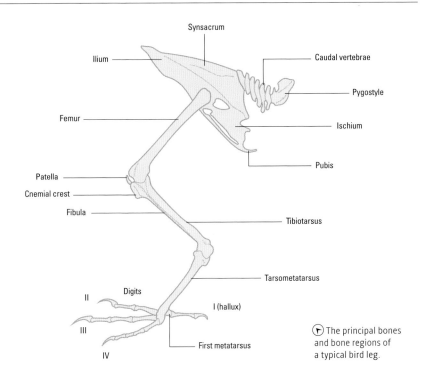

Ⓣ The principal bones and bone regions of a typical bird leg.

The standard toe arrangement is called anisodactyl, but if the basal phalanges of digits 2 and 3 are partially fused (as is the case in kingfishers and some related groups), the foot type is syndactyl. Some birds' feet are arranged differently. In woodpeckers, cuckoos, parrots and some other climbing birds, digits 1 and 4 both point backwards (zygodactyl). Ospreys, turacos and owls can rotate their fourth digit forwards or backwards as required, and so are partially zygodactyl. These alternative toe arrangements have evolved independently in several unrelated groups of birds. Trogons also have two toes pointing forwards and two backwards, but in their case the backwards-pointing digits are 1 and 2 (heterodactyl), and in swifts and some mousebirds, all four toes point forwards (pamprodactyl).

Bird leg bones also include the patella (kneecap), which protects the knee joint; this is, however, absent in some species (and in ostriches, there are two patellae in each knee). There are also a couple of very small, unfused metatarsal bones remaining in the bird foot.

SKELETAL VARIATION

Stripped down to the bone birds' bodies are very similar. The main differences are the relative length of the leg and wing bones, and the shape of the maxilla and mandible in the skull. There are certain adaptations, though, that are peculiar to particular bird families or even species.

The ratites – ostriches and their relatives – are the only birds alive today that do not have a keel on their sternum. This structure, for attachment of large, powerful pectoral muscles, is associated with flight and wing-powered swimming. It is present in other flightless birds, though, as they evolved from flying ancestors. Penguins, which have repurposed their wings into powerful swimming flippers, have a well-developed keel. The bones of their wings are also much shorter, stouter and more rigid and dense than in birds that use their wings for flight.

INGENIOUS ADAPTATIONS

The hyoid bone, a small and slender U-shaped bone in the base of the mandible of most vertebrates, has the function of supporting the tongue. In woodpeckers, it is hugely extended, looping around the back of the skull with its branches coming together in the forehead, just above the bill base. It helps to absorb the vibrations caused when the woodpecker is drumming its bill on wood, protecting the brain (rather like a seat belt) as well as supporting the bird's extraordinarily long tongue.

In some galliform birds, such as turkeys, the tarsometatarsus bears a sharp spur, an outgrowth of bone, which the males use as a fighting weapon. Similar bony outgrowths can be found on the wings of some birds, including screamers, the Spur-winged Goose (*Plectropterus gambensis*), and the Spur-winged Plover (*Vanellus spinosus*). In young birds of the Hoatzin (*Opisthocomus hoazin*), a forest-dwelling bird of South America, the second and third wing 'fingers' are elongated and bear claws, which are used for climbing.

SKULL ASYMMETRY

Owls rely heavily on hearing when hunting, and their skulls show large ear openings behind the orbits. Many owls have asymmetrically positioned ear openings, which give them positionally accurate hearing (because a sound directly above or

(ⴸ) The Ostrich and its relatives have a broad, unkeeled sternum and extremely robust leg and pelvic bones.

(⋀) Hoatzins are the only living birds that have wing claws, a leftover trait from their theropod dinosaur ancestors. By adulthood the claws are lost.

below will reach one ear fractionally before the other). In a few species, this asymmetry extends to the actual anatomy of the side parts of the skull, giving it a lopsided appearance when viewed head-on. The

species with this trait are those with the most nocturnal habits, such as the Tengmalm's Owl (*Aegolius funereus*), which is capable of pinpointing moving prey in near total darkness.

(⤓) The long, delicate hyoid bones of a woodpecker's skull give support for its extremely long tongue. The spongy bone in its frontal skull provides shock absorption.

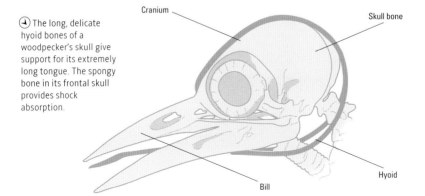

Cranium

Skull bone

Hyoid

Bill

BONES UNDER THE MICROSCOPE

When we think of bones, we tend to imagine them separated from the rest of the body – dried out and long dead. In living animals, though, bone is a dynamic tissue, growing, changing, rebuilding and healing as required.

Hard bone tissue is formed of a tough protein called collagen, mineralised with calcium and phosphorus. However, it also contains several different kinds of living cells, which have functions in maintaining (and changing) the bone tissue. Within the bones is soft, spongy tissue (bone marrow), which generates these and many other kinds of cells, including blood cells. There are four kinds of cells found in bone – osteoblasts, bone-lining cells, osteocytes and osteoclasts. Osteoblasts form new bone, generating the collagen and regulating the deposition of the calcium and phosphorus that mineralise it. Later in their lives, some of the osteoblasts will become flattened bone-lining cells, which regulate the bone's intake and release of calcium. Other osteoblasts become osteocytes, which function rather like nerve cells within the bone tissue, connecting via long branches to other osteocytes. They help direct the activity of the osteoclasts, which dissolve bone tissue. Breaking down bone tissue is as much a part of the growth and healing process as is building new bone.

BONE MARROW

Osteoblasts are large cells made in bone marrow, which sit in dense sheets on the surface of bones. They generate hormones and other products as well as collagen. Most osteoblasts become effectively enclosed by their own products, and at this point they become osteocytes and send out branches

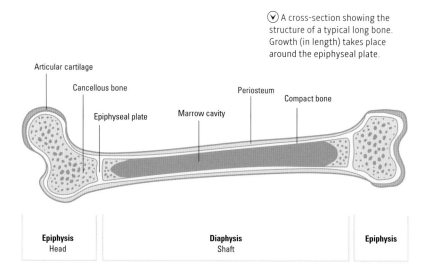

ⓥ A cross-section showing the structure of a typical long bone. Growth (in length) takes place around the epiphyseal plate.

Articular cartilage

Cancellous bone

Epiphyseal plate

Marrow cavity

Periosteum

Compact bone

Epiphysis
Head

Diaphysis
Shaft

Epiphysis

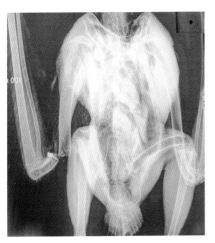

Bone tissue generates continuously throughout life, and bone fractures will heal over time (though not always perfectly).

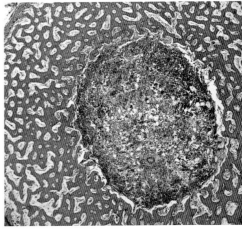

A stained cross-section of an immature bone, showing the bone marrow at its centre, surrounded by bone tissue.

through the bone that communicate with other nearby osteocytes. Those that remain on the bone's surface become bone-lining cells.

Osteoclasts are very large cells with more than one nucleus (occasionally, more than 100). They are formed from aggregations of large white blood cells called monocytes, which are made in bone marrow and have several functions in blood. Osteoclasts are static, though, and anchor into the bone through small projections (microvilli). Through these, they release a hormone called acid phosphatase into the bone, which dissolves both the bone collagen and the minerals it holds.

Bone marrow contains stem cells from which other cell types are made. It is highly dynamic, its composition changing with the individual's state of health, age and other factors. Its main function is the creation of red and white blood cells and platelets, but it also produces the cells from which bone and

cartilage are made. In birds, bone marrow tissue is concentrated inside the ends of the long bones. The parts of the bone that contain marrow are known as cancellous bone, which is spongy, as opposed to the dense and hard cortical bone that forms the outer, rigid structure. Birds' long bones are otherwise hollow, but have some internal reinforcement in the form of narrow struts that cross the inside cavity.

3

THE MUSCLES

Birds are capable of physical feats of which we can only dream. They owe their impressive power and speed to an efficient and extremely hard-working muscular system.

⊙ Birds' large pectoral muscles power flight, but an array of small wing muscles allow the rapid wing-shape changes needed for complex aerial maneuvers.

BIRD MUSCULATURE

As with other vertebrates, birds have a system of skeletal muscles that anchors to the bones and controls the voluntary movement of the various body parts.

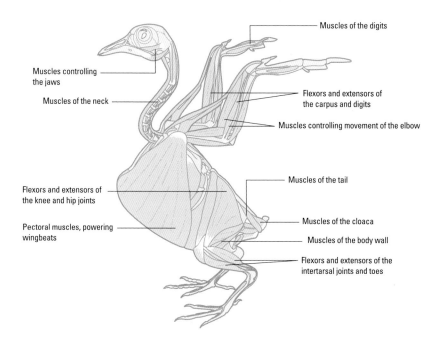

Muscles of the digits

Muscles controlling the jaws

Muscles of the neck

Flexors and extensors of the carpus and digits

Muscles controlling movement of the elbow

Muscles of the tail

Flexors and extensors of the knee and hip joints

Pectoral muscles, powering wingbeats

Muscles of the cloaca

Muscles of the body wall

Flexors and extensors of the intertarsal joints and toes

Ⓐ The general skeletal musculature of a typical bird.

Those of us who eat birds as part of our diet are familiar with the arrangement of their muscles – they form the shape of the bird when feathers and skin are removed. It's obvious from looking at a roast chicken that the largest muscles by far are those in the chest, which attach to the keeled sternum and to the shoulder end of the humerus (upper arm bone). These pectoral muscles make up about 17 per cent of the total body mass (in some cases, such as hummingbirds, up to 25 per cent). They provide the power to flap the wings – within the wings themselves, the muscles are relatively small and control smaller movements.

The thigh and calf muscles are also large, particularly in birds that swim or run. In the Ostrich and other flightless ratites, they are the biggest of all. An Ostrich's hind limb muscles make up about a third of its body weight (to compare to a bipedal mammal, in humans the per centage is closer to 18 per cent). Skeletal muscle is less evident in the extremities but there are many small muscles in the head, neck, wings and feet, controlling all manner of movement. In all, a bird's body contains about 175 individual skeletal muscles.

CONNECTORS

Muscles attach to bones through tendons, the sections of very tough and fibrous collagen tissue at the ends of muscles. The function of a tendon is to transmit the muscle's contraction to the connecting bone, resulting in movement. Muscles only work through contraction; therefore, most joints have sets of muscles working on either side. Contraction of one muscle set straightens the joint, while the other flexes it.

While tendons connect muscles to bones, ligaments are tendon-like structures that connect bones directly to other bones. Cartilage, like muscle, is a connective tissue, but it is firmer and less flexible than muscle – it could be seen as a halfway house between bone and muscle – and is usually found as a cushioning material within joints, helping to protect the bones within from wearing against one another. Bone, muscle, cartilage, ligaments and tendons all contain varying amounts of collagen, a strong and fibrous protein.

ⓥ The massive muscles of an Ostrich's shin help it to power along at a 45-mile-per-hour (70km/h) sprint.

MUSCLES OF THE HEAD AND NECK

Birds may not have very expressive faces, but their heads and necks are highly mobile, and this movement is controlled by an array of muscles. The actions birds perform with their bills are also highly varied and need differing degrees of force and precision.

Just as important as movement is stabilisation – in fast-flight manoeuvres it is still necessary for a bird to use its neck muscles to hold its head steady, so that it is not disoriented by an equally fast-changing view. The opening and closing of a bird's bill is controlled by a set of (usually) seven muscles, which connect the base of the mandible (lower jawbone) to the base of the skull. Most birds have small jaw muscles compared to similar-sized mammals, and this is evident in their skulls – mammal skulls usually have a ridge (sagittal crest) at the back for attachment of large jaw muscles, while the bird skull is smooth at the back.

Parrots have an exceptionally strong bite force, which they use to crush large, hard nuts and seeds. Macaws, the largest of parrots, can exert 1,500 psi (pounds per

Ⓐ An Osprey's (*Pandion haliaetus*) bill is a hooking, tearing weapon but fine muscular control of the jaw also makes it a precision instrument for feeding bites of food to its chicks.

square inch), compared to about 1,000 psi for a tiger, or 1,300 for a gorilla. The parrot jaw contains two muscles that are not present in other birds – the ethmomandibularis and the pseudomasseter. The parrot lower jawbone has a larger, thicker base to allow attachment space for these large muscles.

Ⓡ Parrots can deliver a powerful bite, but their precise muscular control of bill and tongue also allows for delicate manipulations.

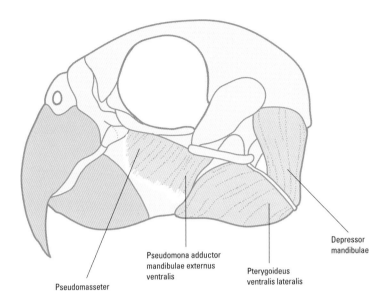

Pseudomasseter

Pseudomona adductor
mandibulae externus
ventralis

Pterygoideus
ventralis lateralis

Depressor
mandibulae

(A) The principal jaw-moving muscles in a
parrot's head. These large muscles give it
tremendous bite force.

THE AVIAN NECK

The neck of a bird contains many more
vertebrae than in mammals, and these bones,
plus an array of mostly slender, long muscles
linking the vertebrae to each other and to the
base of the skull, form a long and highly
flexible S-shaped neck. Because birds do
most of their food-handling with their bills
and not their forelimbs, precise control of the
neck's reach, movement speed and position is
necessary, but the neck also needs to be very
strong to resist the forces it experiences
during flight. Herons, some of which have
very slender necks, usually fly with the neck
folded in to help protect it.

The avian neck is also aided in its freedom
of movement by layers of cartilage on the
top and bottom surfaces of each vertebra,
which allow freely sliding movement. Birds
do not possess the separate cartilage disks
that separate the vertebrae of mammals and
other vertebrates. These adaptations enable
a hunting heron to straighten its neck in
under a quarter of a second to grab a fish in
the water, or a woodpecker to drum its bill
on a tree branch 16 times a second.

FLIGHT MUSCLES

Even birds that are adapted to glide or soar for long periods must still flap to get airborne. For a bird to overcome the pull of gravity on its body mass, a tremendous amount of upwards thrust is needed.

The wing's shape is such that it can push effectively against the air, and also generate lift as it moves forwards, but that is no use without a very powerful downstroke to capture the air in the first place, and this is where muscular strength comes in. The chest of a bird contains a pair of very large pectoralis major muscles, one attached to each side of the sternum's keel, which when contracted bring the wings downward. Their counterparts are the supracoracoideus muscles, located underneath the pectoralis major muscles. When the two supracoracoideus muscles contract, they bring the wings upwards. As the upwards stroke does not generate lift, these muscles are smaller. The pectoralis major is fixed to the shoulder head of the humerus on its underside, while the tendon of the supracoracoideus anchors to the upper side of the same bone to pull it up. This is only achieved because the tendon loops around the top of the furcula (the fused collarbone), creating a pulley system.

When a bird is flying, the contraction of its pectoralis major muscles on each downstroke compresses the air sacs within its chest. Therefore, the bird exhales on the downstroke and inhales on the upstroke.

IN FLIGHT

The muscles in the wing itself are similar to those in our arms – the major ones are in the upper arm and include the biceps to flex the elbow, and the triceps to straighten it. Folding the wing in is important to lose height quickly (for example, when a falcon swoops on its prey, or a gannet dives into water). Holding the wing at full spread is necessary for efficient soaring (using the upwards movement of warm air thermals to gain height without beating the wings), as seen in birds such as storks and eagles.

The bicep and tricep muscles are relatively small, but their presence contributes to the airfoil, lift-generating shape of the wing (thickened on its leading edge). Muscles in

⊙ The wingbeat cycle in birds involves a complex, circular movement, with the wing shape changing dynamically throughout.

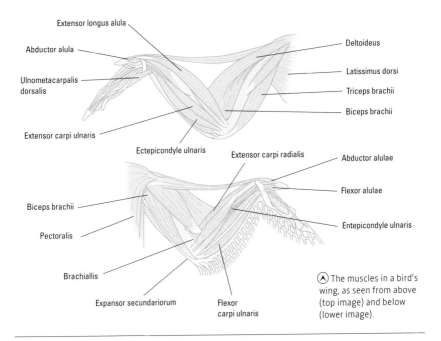

Extensor longus alula

Abductor alula

Ulnometacarpalis dorsalis

Extensor carpi ulnaris

Deltoideus

Latissimus dorsi

Triceps brachii

Biceps brachii

Ectepicondyle ulnaris

Extensor carpi radialis

Abductor alulae

Flexor alulae

Biceps brachii

Pectoralis

Entepicondyle ulnaris

Brachiallis

Expansor secundariorum

Flexor carpi ulnaris

ⓐ The muscles in a bird's wing, as seen from above (top image) and below (lower image).

between the elbow and wrist also affect the flexion of the elbow, and control the position of the digits, including the alula. Extending and fanning out the alula helps the flying bird to delay a stall (loss of lift) when it is flying slowly at a steep angle – the alula forms a small additional airfoil to increase lift. You will often see the alula extended on a bird that is coming in to land on a near-vertical surface.

COMPARATIVE ANIMAL FLIGHT

Powered flight has evolved four times in the animal kingdom – in birds, bats, the prehistoric pterosaurs and winged insects. Even though their wings have very different anatomies, there are many functional similarities because the core requirements for sustained flight are the same in all cases.

When a denser-than-air object is moving forwards through the air, there are four physical forces acting upon it, in two opposing pairs. Weight is the action of gravity pulling it earthward, while lift is the movement of air below it, pushing it upwards. Thrust is the energy driving it forwards, while drag is the air resistance that slows it down. For sustained directional flight, the lift must be stronger than the weight, and the thrust must be stronger than the drag.

Flying animals generate thrust and lift through their wingbeats, enough to overcome their (usually very low for their body size) weight. The wing downstroke, which is (in most flight styles) also a backwards stroke, is performed with the wing fully spread, pushing its maximal surface area against the air. This generates forwards thrust and upwards lift. On the upstroke, the wing is tilted and folded to present a much smaller profile to the air, reducing downward thrust. The animal also makes its shape as stream-lined as possible, for example by tucking up its legs or holding them outstretched behind it. Animals that glide rather than fly, such as flying squirrels, generate some thrust and lift by leaping into the air with their gliding membranes fully extended, but they do not generate any additional thrust or lift by flapping, so their 'flight' is not sustained.

Most flying animals' wings have an airfoil shape – the leading edge is thicker but tapers to a very narrow trailing edge. The most effective airfoils have a curved upper edge and flatter underside edge. This shape, obvious in the wings of aircraft, helps generate lift by reducing air pressure above the wing.

(Ⓐ) Bats' wings are membranous and supported by slim and greatly elongated 'fingers'.

(Ⓐ) Hummingbirds beat their wings more rapidly than any other bird, so their energy expenditure is higher.

Insect flight

Insect flight works in a more complex way than that of the flying vertebrates. The wings move through their stroke in a circular motion, sweeping forwards first, and then backwards with the entire wing twisting over during the stroke so the leading edge points backwards. The forwards movement creates a vortex of low pressure above the wing, and the backwards movement generates backspin. Both actions provide lift to the wing, and the vortex wake from each previous wingbeat provides additional lift when the wing is momentarily stationary between wingbeats.

Most flying insects have four wings, and in cases where the wings are equal in size, the two pairs' strokes can be timed differently, which allows for more efficient hovering than bats or birds can achieve. Dragonflies are perhaps the most adept of all flying animals, able to hover and fly backwards as well as reach straight-line speeds of more than 25mph (40km/h), and to capture all kinds of other flying insects on the wing.

Some small insects fly by clapping their wings together above their backs, then throwing them forcefully apart to create a vortex over each wing that sucks the insect upwards. This flight method – clap and fling – causes heavy wear on the wings, but the insects that use it, such as thrips, have very brief lifespans.

(∧) Dragonflies and most other insects have two pairs of wings that move independently.

(∧) The pterosaurs' membranous wings were long and narrow, supported to their tips by the elongated fourth digit.

LOWER LIMBS

Although most birds undertake longer journeys on the wing, their legs and feet are in constant use daily, and for many species, leg power is the primary means of propulsion. The fastest two-legged runner on Earth is a bird (the Ostrich), and most birds that swim are leg- and foot-propelled.

The many small and large muscles in a bird's lower limbs are not dissimilar to those in the mammalian hind leg. Hip extensors and flexors attach between the femur and the pelvis, controlling movements of the hip joint, while muscles between the tibiotarsus and the femur control flexion and extension of the knee joint. The largest muscle in the leg of a bird built for running, such as the Emu (*Dromaius novaehollandiae*), is the gastrocnemius, which attaches from the

⊙ The Secretary Bird uses its long, strong legs to chase prey and kick it to death.

knee to the ankle and points the foot downward when it contracts. This muscular action is the power-generating part of a running or leaping motion, and also of a swimming kick.

Fast-running birds have long legs for their size, to allow for a greater stride length, but the bulky parts of the leg musculature are concentrated close to the body, making the more distal parts of the leg thin and light and so able to swing more quickly through the running stroke. The Ostrich can run at about 45mph (70km/h) and the Greater Roadrunner (*Geococcyx californianus*), which weighs only about 300g (10oz), at 20mph (32km/h).

THE BIRD FOOT

In many cases, birds' feet are capable of precise, powerful, or delicate movement. Birds of prey often strike their prey with their feet, and the force they generate with this forwards thrust can be considerable, though in most cases the force is increased because the bird is dropping down through the air before the strike. The Secretary Bird (*Sagittarius serpentarius*), which hunts on foot and kills prey (including venomous snakes) with powerful kicks of its long legs, is an example of a species that strikes at prey with leg power alone. It kicks with a force more than five times its own body weight. The impact is extremely brief (lasting only about 15 milliseconds before the foot is withdrawn), enabling the bird to avoid a bite from its prey if its kick misses the target.

Birds have specialised flexor tendons in the leg and foot that connect the undersides of the toes to the ankle and knee joints in a pulley arrangement. When the leg is bent at these joints, the tendons pull on the toes and make them curl up. A bird going to roost on a branch crouches to activate this reflex, and the toes will not release until the bird straightens its legs, ready to take flight. The same process causes a bird of prey to close its talons when it attacks – as it pounces on its prey, its weight moving forwards bends the joints of the leg and tightens the flexor tendons. Some birds hold their legs folded when they fly, and if they pass overhead you can see that their toes are curled up. Others, such as cranes, fly with their legs extended and you will notice that their toes are held straight.

Birds that swim on the water's surface use powerful alternating backwards kicks to propel themselves. They can control their speed and position precisely, as demonstrated by phalaropes, which spin rapidly on the water's surface to stir up small aquatic prey.

Ⓐ The Greater Roadrunner, a type of cuckoo, is one of the fastest-running of the flying birds.

Ⓥ Penguins use their webbed feet to swim at the water's surface, but underwater they are mainly wing-propelled.

OTHER MUSCLES

The bird's primary muscles are those that control large-scale movement of its limbs, neck and head, but there are many more. They have a role in both voluntary and involuntary movement, from changing the size of the pupil to controlling which sounds a singing bird produces.

Birds do not have a diaphragm – the sheet of muscle that sits below our lungs and contracts when we breathe in. When we exhale, it relaxes. In birds, exhalation is the active phase of respiration. They inhale passively through relaxation of the intercostal muscles (small muscles between the ribs), which pushes the ribs and sternum outwards. When the intercostal muscles contract, the bird exhales.

The bird's tail, although it lacks bony support except at its base, functions rather like a fifth limb and has its own dedicated musculature, which functions mostly separately from the muscles controlling wing and leg movement. The main muscles involved are the lateralis caudae and the levator caudae. These caudal muscles allow the long tail

⊙ Muscles in the skin control the raising and fanning out of feather groups, as in the dramatic display of the male Peacock (*Pavo cristatus*).

feathers to tilt, twist and fan out or pull together. Changing the shape and orientation of the tail allows the bird to control its speed and direction in flight.

OTHER MUSCLE TYPES

Other feathers and feather groups (tracts) are also under muscular control. Some of the small skin muscles that raise or lower feathers are made from a different kind of muscle tissue than the striated tissue that constitutes skeletal muscles. This is smooth muscle. Like striated muscle, it contains actin and myosin filaments, which move over each other to perform muscular contraction, but the filaments are not orderly, and smooth muscle contraction is involuntary.

Birds' feathers may be raised or lowered involuntarily in response to changes in temperature (fluffed-up feathers help trap heat, flattened feathers release it). Voluntary feather movement, controlled by small striated muscles, is also commonplace, and can be seen in spectacular fashion in some of the courtship displays performed by a variety of bird species. Male Sage Grouse, peacocks and many other gamebirds raise and fan out their tail feathers, while males of the various species of birds-of-paradise erect and quiver different sections of contrastingly coloured and curiously shaped body plumage to impress females.

Smooth muscle tissue can also be found within the blood vessel walls, the gastrointestinal tract, in the eye (controlling the shape of the iris and lens), and inside certain internal organs. The walls of the digestive tract, for example, contain smooth muscle in places, which contracts to push food along or to break it down. The muscle that forms the walls of the heart is a third

(A) Muscular control of feather position is very important in body temperature regulation.

kind of tissue, cardiac muscle. It is a kind of striated muscle tissue, similar to that found in skeletal muscles, but its contraction is involuntary.

(>) The courtship display of the male Sage Grouse features a fanned tail and inflated sacs at the front of the bird.

SWIMMING AND DIVING

The surface and depths of water offer plenty for an enterprising bird to eat, and it's not surprising that the ability to swim, dive, or do both has evolved in a range of unrelated bird families.

For a bird to be at ease on or in the water, and remain capable of flight, requires a range of adaptations, though penguins and some ducks and cormorants have indeed become flightless, and evolved as better swimmers as a result. Surface swimming is not too much of a leap for a land bird, as birds' bodies have a low enough density (thanks to air within the air sacs and trapped in their feathers) that they will naturally float on water. In most cases their plumage also has very good natural waterproofing. All the bird then needs to do is hold its balance and propel itself with kicks of its feet. Even land birds that fall into water can usually swim to safety before becoming waterlogged, and waterside species such as rails, sandpipers and herons can swim short distances with reasonable competence.

Birds that are habitual surface swimmers have webbing between their toes, making the foot a more efficient paddle, able to push against a bigger volume of water to drive forwards. They also have relatively short, powerful legs, often set towards the rear of their rather elongated, boat-shaped bodies. The non-diving species, including swans, geese, moorhens and gulls, will pick food from the surface, but they also usually have long necks, to reach underwater when they submerge their heads.

Most birds that swim are capable of diving as an emergency measure at least, even if they do not need to dive to find their food.

ⓥ The Mandarin, a dabbling duck, has a more buoyant swimming position than duck species that feed by diving.

58

Diving

When it comes to diving, some birds dive from the water's surface while others plunge in from the air. Many plunge-divers are not swimmers, and rely on making a shallow dive and using powerful wingbeats to lift clear of the water once they have caught their prey. Such birds usually have broad wings, to generate maximum lift as quickly as possible. Gannets and boobies plunge-dive from a considerable height. To protect themselves when entering the water at speed they have sealable nostrils that open inside the bill rather than externally, and large air sacs, plus a very sturdy sternum to cushion the impact of the water on their bodies. Most birds that swim underwater, and especially those that hunt in cold waters, such as the Antarctic penguins, keep their bodies warm and dry thanks to very smooth, dense plumage that keeps water from reaching the skin, instead trapping a layer of air there.

Underwater swimming is hard work – the bird must overcome its own natural buoyancy, which only occurs when it gets to a depth where the water's density is the same as that of its own body. Birds like penguins and auks must dive very deep (often more than 30 metres/100 feet) before they can easily begin to chase fish underwater. Cormorants and shags are much less buoyant than most diving birds because their plumage is not waterproof, so no air is trapped under the feathers. They can therefore pursue prey at shallower depths, but have to dry in the breeze and sunshine after each dive.

ⓐ By pulling its wings back, a diving Gannet transforms its body into a streamlined javelin shape, enabling it to dive deeper.

ⓥ Lacking external nostrils ensures that a Gannet doesn't take in unwanted water and risk drowning when it hits the surface.

MUSCLE MICROSTRUCTURE AND FUNCTION

Skeletal muscles are made of striated muscle tissue, a form of connective tissue that can change its shape (contract) to a significant degree. Striated muscle tissue has a unique microstructure and chemical composition that allows this.

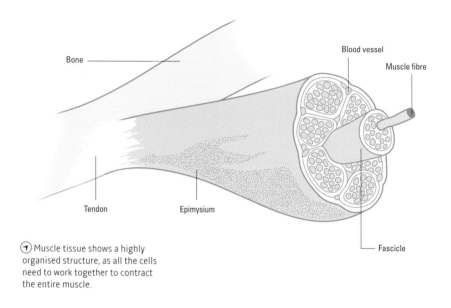

Bone

Blood vessel

Muscle fibre

Tendon

Epimysium

Fascicle

(↑) Muscle tissue shows a highly organised structure, as all the cells need to work together to contract the entire muscle.

The muscle cell is also known as a myocyte, or a muscle fibre. It is tubular in shape and contains two kinds of protein filaments – actin and myosin. These are arranged in overlapping groups, with the tips of the thick myosin filaments separated by a space called the Z-disc, and the tips of the narrower actin filaments by a space called the H-zone.

Each section of filaments between two Z-discs is called a sarcomere, and many sarcomeres are joined end-to-end to make each muscle cell. This gives the cell a banded appearance, with the Z-discs forming evenly spaced dark lines. When contraction occurs, the actin filaments slide over the myosin filaments to narrow the H-zone, and the cell shortens as a result, though its Z-discs are unchanged in appearance, just closer together.

Muscle cells receive stimulation from motor nerve cells, which send signals to them to contract when the animal 'decides' to move the body part in question – the process is nearly instantaneous.

A CHEMICAL REACTION

When a nerve impulse stimulates a muscle cell, sodium ions enter the cell and calcium ions leave it. This chemical change causes the bonds holding the actin and myosin filaments in place to break down, allowing them to move closer together. In this way, the cell shortens, or contracts.

Within a muscle, the muscle cells are organised into bundles called fascicles, and several of these are grouped together to form the complete muscle. Blood vessels run between the fascicles, branching off into capillaries which supply individual cells with blood.

The muscle is entirely encased in a thick, tough membrane, made primarily from collagen. This membrane is called an epimysium, and it extends beyond the soft interior of the muscle to form the tendons, which attach to bone at either end (or, in some cases, to the skin). The muscle tissue itself varies in colour from reddish to much paler. Red muscle is known as 'slow twitch'. It carries fewer muscle fibres but has a richer blood supply, bringing enough oxygen for efficient aerobic functioning. Paler muscle is 'fast twitch' – it has more muscle fibres so contracts quickly, but its lower blood supply means it works anaerobically (without oxygen). This is less efficient, so the muscle tires quickly.

(ⱽ) For juvenile eagles battling for scarce resources, it is vital to be able to move at high speed and with commitment, using the fast-twitch muscles.

THE NERVOUS SYSTEM

Birds' brains are structured very differently to those of humans and other mammals, but studies increasingly demonstrate that avian intelligence and responsiveness is far more advanced than we used to believe.

⊙ High-speed nerve pathways between the eyes and the brain allow a Golden Eagle (*Aquila chrysaetos*) to make lightning-fast decisions as it hunts.

THE BIRD NERVOUS SYSTEM

Birds live in a fast-paced world, and need a highly refined nervous system to keep up. The nervous system is concerned with how a bird channels and organises the information about the outside world from its senses, and how it responds to that information.

The nervous system has a key role in regulating various internal physiological processes, as well as higher mental functions. Like other tetrapods, birds possess a well-developed brain and a spinal cord (the central nervous system). The spinal cord comprises numerous bundles of very long, fibre-like nerve cells or neurons. The brain is protected within the skull, and the spinal cord runs through the central holes of the chain of vertebrae, sending nerves out to other body parts on the way, forming the peripheral nervous system. The larger nerves comprise multiple neurons but they continue to branch, each offshoot becoming smaller. Individual nerve cells reach into virtually all body tissues, and they all connect to other neurons and eventually to the brain. Some neurons in the body are afferent, carrying signals to the brain from the destination tissue, while others are efferent, carrying signals in the other direction.

The brain itself contains a vast number of neurons, each with multiple connections with its neighbouring neurons. Neurons in the brain and the peripheral nervous system are well protected by other tissue types.

ACTIVE AND AUTOMATIC

The nervous system's activity guides a bird's behaviour, from simple to complex actions. Nerve impulses convey sensory information

ⓐ The European Bee-Eater (*Merops apiaster*) needs quick reactions and precision timing to capture a bee and avoid being stung.

to the brain, which processes this information and decides upon a response, and then impulses travel along efferent neurons to carry the brain's instructions to the skeletal muscles, triggering them to contract. This cycle happens at extraordinary speed, enabling a flycatcher to seize an insect on the wing, a snipe to dodge out of the path of a pursuing hawk, and a fast-flying flock of shorebirds to coordinate their movements in a dazzling switchback display.

The nervous system is intimately involved in all other bodily systems, too. It plays a key role in regulating the release of hormones and enzymes, and manages the beating of the heart, the reproductive cycle, and the inner workings of the digestive system.

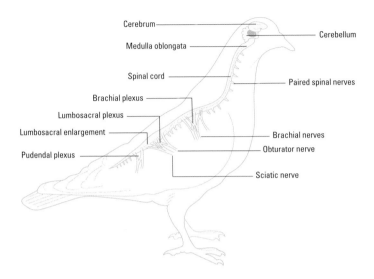

Cerebrum

Cerebellum

Medulla oblongata

Spinal cord

Paired spinal nerves

Brachial plexus

Lumbosacral plexus

Lumbosacral enlargement

Brachial nerves

Pudendal plexus

Obturator nerve

Sciatic nerve

ⓥ Starlings flying together in beautiful synchrony give the impression of being controlled by a unified intelligence.

ⓐ The brain, spinal cord and primary nerve pathways of a pigeon.

TEMPERATURE CONTROL

The ability to regulate the body's internal temperature is what enables birds, and mammals, to live and be active in very cold conditions, as well as hot, and thus colonise almost everywhere on Earth.

Birds and mammals are the only endothermic animals (able to internally regulate their own body temperatures) – all others rely on the ambient temperature warming their bodies to a point where they can function. Homeostasis, the body's various systems and methods for maintaining a balanced internal environment, involves interactions between many different organs, from the skin inwards. Maintaining the right internal temperature begins with 'knowing' what that temperature should be, a process overseen by the hypothalamus, a structure at the front of the brain. When a bird's core temperature increases, the hypothalamus triggers cooling processes, such as panting, and vasodilation (dilation of the blood vessels). Heat is more readily lost through bare skin than feathered skin, and accordingly some birds have vascularised bare throat pouches, which they can flutter in a breeze to lose heat easily.

Internal cooling also stimulates the hypothalamus, and triggers heat-conserving physiological processes such as vasoconstriction, and fluffing up the plumage to trap air and warm it against the skin. Temperature-raising behaviours include perching on one leg to keep the other tucked warmly in the feathers, and seeking out fat-rich foods to provide extra fuel.

The optimum human body temperature is 37°C (98.6°F). Most birds maintain a warmer internal temperature than this, with most species staying at 39–40°C (102–104°F), regardless of size and habitat. Birds from cold climates have thicker, denser body plumage

ⓐ Resting on one leg at a time allows the Grey Heron (*Ardea cinerea*) to keep the other foot warm within its belly plumage.

and sometimes have feathered feet (these are also characteristics of swifts that fly at high altitudes in cold air), while birds that swim in cold water have exceptionally good waterproofing in their feather structure, to keep their skin dry when they submerge. Birds dwelling in hot countries, which lose heat through water evaporation from their mouths when they pant, depend on easy access to a reliable water source to replenish this lost fluid.

A state of torpor

For tiny mountain-dwelling hummingbirds, with their extremely high energy requirements, maintaining that 40°C (104°F) temperature through the cold night when they cannot feed is more than they can manage. Instead, they enter a state of torpor, and reduce their energy needs to as little as 5 per cent that of when they are awake. The internal body temperature falls to about 18°C (64°F), and heart and breathing rates drop dramatically. In this inactive state, the hummingbird is unable to react to any threats, so must choose the safest possible roosting place before becoming torpid. Some other birds also experience torpor. However, no bird enters true hibernation as mammals do, spending the entire winter torpid to survive months of cold conditions with little food around. The preferred avian strategy is to migrate to more hospitable climes. Interestingly, in bats, which are as strong-flying as many migratory birds, hibernation is a more frequent tactic than migration.

ⓐ Bathing can help a bird to cool down on a hot day, though the feathers' waterproofing makes this less effective than it is for a furry mammal.

ⓥ Many species of bats survive freezing temperatures by becoming torpid, but this strategy is rare in birds.

BRAINS: SIZE AND INTELLIGENCE

The expression 'bird brain' betrays our lack of respect for avian intelligence. Some birds are indeed not overburdened with cleverness, but studies show that the brainpower of corvids (the crow family) in particular is on a par with that of the most intelligent mammals.

We can get some indication of how intelligent an animal is by comparing the size of its brain relative to the total size of the animal. Brain mass compared to total body mass can be expressed as a ratio. Among birds, the brain:body mass ratio is highest in intelligent species such as crows and parrots, and in humans the ratio is much higher (1:40) than in our close cousin the chimpanzee (1:113). However, small animals always have bigger brains relative to body size than larger ones – for example, the brain:body mass ratio for the mouse is the same as for a human. This effect is corrected in a more refined way of measuring brain size – the encephalisation quotient (EQ). Under this system, the human EQ is 8.1, higher than any other mammal, while the chimp's EQ is 4.2 and the mouse's 0.5.

Birds fare similarly to mammals in EQ. However, overall brain size relative to body size is not the whole story when it comes to cognitive processing power. The total number of neurons within the brain is of more importance. Birds have more neurons packed into their brains than mammals do – a space- and weight-saving adaptation. Ravens and macaws have a higher neuron count than many much larger and clearly intelligent mammals, such as dogs and raccoons. Also important are the interrelationships between those neurons – how many connections they have, and how quickly nerve impulses are conducted along the neuron and across from one neuron to the next. Birds score highly in these areas as well.

ⓒ Chimpanzees are among the most intelligent of mammals, but birds such as ravens are their intellectual equals.

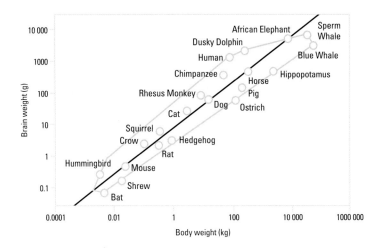

(A) Brain size relative to body weight for a range of mammals and birds.

HOW THE BIRD BRAIN WORKS

Not all brain regions are concerned with actual information-processing, adaptation, abstract thought and problem-solving – the standard measures of intelligence. In mammals, the part of the brain that deals with these higher or executive functions is the neocortex. This is the six-layered, highly organised frontal part of the brain, and is derived from the pallium, or outer layer of the brain. The neocortex is larger in more intelligent species. The pallium of birds has not given rise to an organised neocortex, which is why they historically were presumed to have low intelligence. However, the pallium has a much higher density of neurons than the mammalian neocortex, and includes specific regions that carry out the executive functions in the same way as the mammalian neocortex.

Tests of intelligence at which corvids excel include making and using tools to solve complex problems, recognising themselves in mirrors, and planning for anticipated needs – tasks that are far beyond the capabilities of many much larger-brained mammals. Of course, every bird and mammal is adapted to a particular lifestyle, and those adaptations include variations in the relative size of different brain regions. For example, dolphins' brains contain a particularly well-developed paralimbic region compared to land mammals, which may give them a greatly improved speed of perception.

BIRD BRAIN STRUCTURES AND REGIONS

Until recently, the study of birds' brains has been cursory, and there are still some areas of the avian brain the function of which remains somewhat mysterious. However, we do know that different brain regions control different functions within the body.

Birds' brains have three general regions – the hindbrain, midbrain and forebrain. The hindbrain contains most of the brain stem (where the brain transitions into the spinal cord) and the medulla oblongata. This area's nervous input and output is mainly concerned with automatic bodily processes, such as controlling the bird's heart rate and breathing, and regulating the blood pressure. It is also involved in involuntary reflexes.

Another structure usually considered to be part of the hindbrain is the cerebellum, a distinct, convoluted structure concerned with coordinating physical movements. It also receives input from the middle ear about the head's position in space, so helps to maintain balance. It is relatively large in birds, which

reflects the high-speed and highly coordinated movements that most birds have to make in flight.

The two halves of the midbrain each include a large optic lobe, containing layered neurons, which receives input from the optic nerve, and contains neurons that deal with processing visual information. By contrast, the olfactory bulb at the front of the brain, concerned with perceiving odours, is proportionately very small in most birds. However, it is well-developed in certain seabirds and scavengers.

ⓥ The main structures and regions of the avian brain.

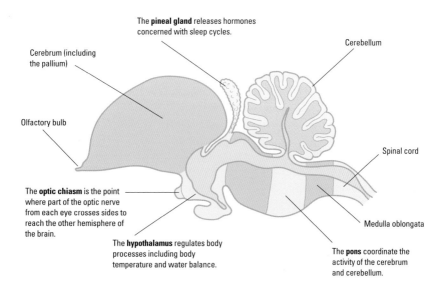

The **pineal gland** releases hormones concerned with sleep cycles.

Cerebellum

Cerebrum (including the pallium)

Olfactory bulb

Spinal cord

The **optic chiasm** is the point where part of the optic nerve from each eye crosses sides to reach the other hemisphere of the brain.

Medulla oblongata

The **hypothalamus** regulates body processes including body temperature and water balance.

The **pons** coordinate the activity of the cerebrum and cerebellum.

The pallium, forming the outer layer of the two halves of the forebrain, includes several distinct regions. The nidopallium is situated in the forebrain and is thought to be involved in higher cognitive function in birds – learning, innovating and remembering, for example. Other brain regions include several linked clusters of neurons involved in learning and producing song. The density of neurons per gram of brain is very high in small birds in particular; the Common Starling (*Sturnus vulgaris*), for example, has 260 neurons per gram (7,370 per ounce) of brain tissue, compared to just 57 neurons per gram (1,616 per ounce) in a human.

As well as neurons, birds' brains contain other kinds of cells, in particular various

ⓐ During high-speed coordinated courtship displays, the cerebellum's activity reaches a peak.

types of glial cells, which provide physical support and protection for the neurons, produce the cerebrospinal fluid that cushions the brain, and remove waste products. However, glial cells are much less abundant in the brains of birds compared with mammals'.

ⓧ Ravens are highly intelligent birds, capable of abstract thought and complex problem-solving.

SOCIAL BEHAVIOUR

Being social is of great benefit to birds in some circumstances, and many species live permanently in social groups, foraging and breeding together. In long-standing social groups, it is even possible for deeper, lastingly beneficial bonds – friendships – to develop between individuals.

Sociality means more eyes on the lookout – for feeding opportunities and for danger. It also gives each individual more chance of finding a suitable mate, and a reduced chance of being the target if a predator does attack the flock. In cold weather, roosting in groups helps conserve and concentrate heat. During migration, travelling in flocks helps keep all individuals on the right track.

For effective social behaviour, birds need to have good social signalling. This involves both vocal and visual communication. Flocks on the move tend to call constantly. These usually brief but far-carrying contact calls help each bird ensure that it remains close to the main part of the flock and help draw in any lone birds – they are especially valuable in night-migrating species. Birds that travel in flocks also often have distinctive markings that are revealed when the wings are spread. These include the speculums of ducks – panels of white-bordered iridescent colour in the secondary feathers, present even in the otherwise drab-plumaged females. These markings are striking from a distance, even in low light. Migrating flocks tend to be led by the fastest fliers, but if these birds prove themselves to be less able as navigators, the flock learns to ignore them in favour of slower but more capable alternative leaders.

Social gatherings are not necessarily egalitarian. Some birds consistently

⌄ Migrating in flocks keeps Snow Geese safer from predators, and young birds benefit from the opportunity to learn from their elders and meet potential mates.

⌃ Large birds travel in a V-formation to enable each (except the leader) to use the slipstream of the bird in front.

command the best positions, typically near the centre of the flock, whether for feeding, roosting, or travelling. The most dominant individuals in a social group tend to be the strongest and are likely to be the same birds that fare best in more directly competitive situations. Eurasian Wrens, which are aggressively territorial, will form tightly packed communal winter roosts to share their body heat, inside tree holes or other cavities. The bird that holds the territory in which the roost forms will attract others to the roost with calls, but birds that have travelled a long distance to the roost are driven away – probably because they pose more threat to the territory-holder than its settled, territory-holding neighbours.

Beneficial tolerance

Some social links are uneasy, to say the least. Male Ruffs (*Calidris pugnax*) develop elaborate head ruffs and gather on leks (display grounds) to attract females in spring. The darker-ruffed males battle each other, but they tolerate the presence of non-displaying white-ruffed 'satellite' males as they draw extra female attention. Both classes of males will have opportunities to mate, as will a third type, the 'faeder', which has no male ornamentations and appears to go unnoticed by the other two, but not by females. Studies show that the three types of male Ruffs are highly distinct genetically as well as behaviourally, and all are perpetuated by female choices.

Avian friendship

Work on Great Tits (*Parus major*) in England has shown that individuals in winter-feeding flocks tend to spend more of their time with certain birds than others. It was also noted that the most aggressive and assertive birds had more social links but that they were less stable, while timid birds had fewer but longer-lasting bonds. The benefits of such friendships could include a growing familiarity with each other's particular behaviours in response to important things like food sources and possible approaching predators.

COMPARING BRAINS

In birds, brain size and structure vary considerably from species to species, and in some cases, they even vary somewhat between the sexes. The relative sizes of the brain regions are indicative of the lifestyle and the dominant senses for that particular species.

Brain size is linked to body size but also to intelligence, and as discussed on page 68, the birds with the highest brain size relative to their bodies, and also the most well-developed forebrains, are the crows and the parrots. These species, in general, are long-lived and sedentary birds, and both show a remarkable capacity to learn and to innovate. Parrots are primarily vegetarian, and within their habitats they will eat different kinds of fruits, seeds and foliage at different times of year. This means they need to learn and remember when and where each kind will be available, and the best way to obtain it. Crows are opportunists, able to take food of many different kinds as long as they can work out how. Memory and innovation are also vital for them to survive and thrive.

(A) Migrating Barn Swallows (*Hirundo rustica*) always return to the same place to breed, using memories of local landmarks to find their way.

BRAIN ADAPTATIONS

The optic lobes are large in most birds, but particularly in those that are long-distance migrants, suggesting that visual information may help them to navigate and to quickly find the resources they need in many different areas. However, these birds generally have smaller brains overall than non-migratory species. Sedentary birds build a detailed knowledge of their habitat and territory over time, to help ensure they can

(<) The mysterious 'Area X' in the brains of male songbirds like Zebra Finches helps them to learn their species-specific song.

> Macaws learn the layout of their forest home and visit each tree at the time when its fruits are ready to eat.

find food even in times of severe scarcity. This is a capacity that long-distance migrants do not need so much, as they make seasonal journeys to ensure they can find good food supplies all year round. However, although migrants use non-learned cues (such as Earth's magnetic field) to broadly direct their migration paths, they use their memories of particular details to home in on favourite stop-off points along the way.

The olfactory bulb is small in most birds but is very large in some that have a well-developed sense of smell, such as the petrels, rails and kiwis. The optic lobes of kiwis are much reduced, showing how smell has replaced sight as a dominant sense in these mainly nocturnal ground foragers.

Male Zebra Finches (*Taeniopygia guttata*) and some other songbirds have a specialised brain region known as 'Area X', which is one of several involved in learning and copying sounds. These birds do not instinctively 'know' their species' courtship song and have to learn it by listening to other males. Females do not learn to sing themselves, but have other brain regions that allow them to learn and recognise the male's song.

∧ Birds that forage on the ground in marshy, muddy habitats, such as this Little Crake, usually have a good sense of smell.

THE PERIPHERAL NERVOUS SYSTEM

Nerves branch off in pairs from the brain and spinal cord and eventually penetrate all the body's tissues. These nerves form the peripheral nervous system, arising from the central nervous system (the brain and spinal cord).

The nerves that originate from the brain itself are cranial nerves, while those that branch off farther down the spinal cord are spinal nerves. Birds, like other higher vertebrates, have 12 pairs of cranial nerves, some wholly afferent, some efferent, and some containing fibres involved in both. In most cases, their names indicate their primary function. Eight of these nerves come from the hindbrain. The remaining four are: 1) the olfactory nerve, connecting the olfactory bulb to scent receptors in the nostrils; 2) the optic nerve, linking the retinas of the eyes to the optic lobes; 3) the oculomotor nerve, connecting the midbrain to some of the muscles controlling eye movement; and 4) the small trochlear nerve, which also controls eye movement.

(A) The Great Grey Owl (*Strix nebulosa*) has exceptional hearing. Sound inputs from the ear reach the brain via the vestibulocochlear nerve.

The other eight cranial nerves, between them, serve the muscles controlling the bill and tongue, receive input from the hearing and balance-regulating parts of the ear, receive sensory input from the taste buds in the tongue, regulate facial glands, and control the muscles involved with vocal activity.

One of them, the vagus nerve, is very long and branches out to the rest of the body. It is a key part of the autonomic nervous system. This has two components – the sympathetic and parasympathetic nervous systems. We feel the action of our own sympathetic nervous system when we receive a shock – its

(>) Neurons have a branching structure, which allows them to communicate with many other nearby neurons at the same time.

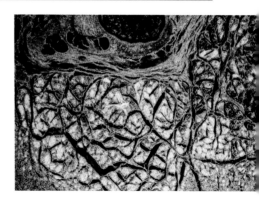

effects include a rapid rise in heart rate, pupil dilation, a slowing of stomach activity, and other physical adjustments to ready our body for a 'fight or flight' response. The parasympathetic nervous system triggers the opposite response, when the danger has passed ('rest and digest').

THE SPINAL NERVES

The spinal nerves regulate processes in the body's organs and muscles, and receive sensory input from them. They emerge from holes in the vertebrae, and each contains efferent and afferent fibres. As they branch and divide, eventually they send individual neurons into the tissues.

Nerves handle all of a bird's sensory and motor functions. If damaged, nerve tissue is slow to regenerate, but shows the ability to form new connections to compensate.

(v) The main pathways of the cranial and upper spinal nerves in the chicken head and neck.

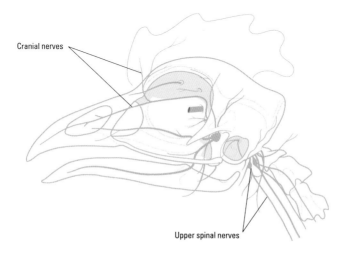

Cranial nerves

Upper spinal nerves

NERVES AND NEURONS

Neurons are among the most specialised and distinctive cell types in the body. Neurons within the brain differ from those within peripheral nerves, mainly in their length. Those in nerves are much longer. However, the way they work is similar in both locations.

The structure of a major nerve is quite similar to that of a muscle, with the neurons organised into several bundles (called fascicles, as are the bundles of striated muscle cells within a skeletal muscle). Each fascicle is contained by a membrane called a perineurium. Blood vessels run between the fascicles, and the entire nerve is protected by a sturdy membrane of connective tissue – the epineurium.

The neuron itself is shaped a little like a tree, with a branching cell body (soma). From the soma, a long slender stem (axon) extends, eventually forming a smaller collection of shorter branches at its tip. The soma contains the cell's nucleus, and its many fine branches (dendrites) receive signals from other neurons, which they transmit to the axon.

Depending on their location in the body, neurons can have immensely long axons (some in the human body are over one metre [three feet] long). The nerve impulse is the electrical signal that travels along the axon, and its progress is speeded up because of the fatty myelin sheaths that surround the axon. The myelin is a poor conductor and works like insulation around a wire, so the signal 'jumps' along the spaces between the sheaths.

SYNAPSES

Signals travel from neuron to neuron across the small space (the synapse) between the axon terminals of the first cell and the dendrites of the next. There are two kinds of synapse. The first, electrical synapses, in which the electrical impulse crosses the gap directly, work very quickly. The second

⊕ Neurotransmitter molecules released from the terminal branches of one neuron bind to receptor molecules on the neighbouring neuron's dendrites, with enzyme action accelerating the reaction.

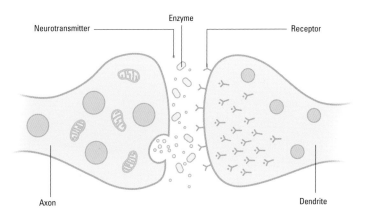

Neurotransmitter — Enzyme — Receptor

Axon

Dendrite

kind is the chemical synapse. In these cases, when the nerve impulse reaches the axon terminal, the terminal releases a particular chemical called a neurotransmitter. These molecules cross the synapse and fit into special receptors in the cell membrane of the dendrite of the receiving neuron.

Ⓐ Recognising and responding to social cues requires a very rapid nerve signalling process.

Their arrival sets up an electrical charge in the dendrite, which passes to the soma and on into the axon.

Ⓓ The structure of an individual neuron or nerve cell.

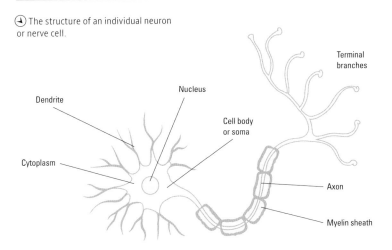

Dendrite

Nucleus

Cytoplasm

Cell body or soma

Terminal branches

Axon

Myelin sheath

THE
SENSES

All living things sense the world around them in one way or another. Most birds are gifted with exceptional eyesight and hearing, backed up by various other senses, some familiar and some strange to us.

> A Long-eared Owl's eyes are as prominent as they are beautiful. The 'ears' on top of its head are feather tufts. Its real ears are hidden under the feathers on the sides of its head.

EYE ANATOMY

Nearly all birds rely on vision above all other senses to guide them around their world. Because they are so fast-moving, they need a visual system that works extremely rapidly and accurately.

Avian eyesight is rightly famous for these qualities – the visual abilities of birds of prey, for example, are the stuff of legend. However, in essence the bird eye is not radically different from the eyes of mammals like us. The avian eyeball is protected by a thick white membrane (sclera), except at the part that is open to the air. This is covered by the transparent and slightly bulging cornea, and where the cornea graduates into the sclera are the small, supporting bones of the sclerotic eye-ring (see page 35). The cornea sits above the iris, a coloured, circular muscle with a central hole (pupil) that enlarges and shrinks as the iris contracts, changing the amount of light that can pass through. Light passes through a transparent lens suspended just behind the iris – this focuses light on to the retina, which lines the inside wall at the back of the eyeball. Its light-sensing cells are connected, via neurons, to the optic nerve, which exits at the back of the eyeball. The optic nerves from each eye connect to the optic in the brain (see page 70).

INSIDE THE EYEBALL

The interior of the eyeball is filled with a gel-like substance called vitreous humour, while the space between the cornea and the iris contains the clear and more fluid aqueous humour. Between the light-sensing retina and the tough, protective sclera lies another layer of tissue, the choroid, which carries blood vessels supplying the eye. An extension of the choroid, the pecten, is a concentrated section of blood vessels, which projects into the eyeball and sits above the optic nerve. The pecten's blood vessels convey nutrients and help maintain the correct pH balance of the vitreous humour. By packing many blood vessels into a small space, the disruption they cause to the path of light to the retina is kept

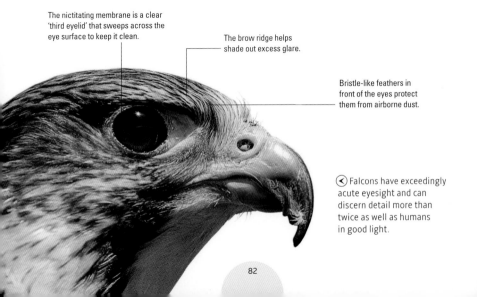

The nictitating membrane is a clear 'third eyelid' that sweeps across the eye surface to keep it clean.

The brow ridge helps shade out excess glare.

Bristle-like feathers in front of the eyes protect them from airborne dust.

⊘ Falcons have exceedingly acute eyesight and can discern detail more than twice as well as humans in good light.

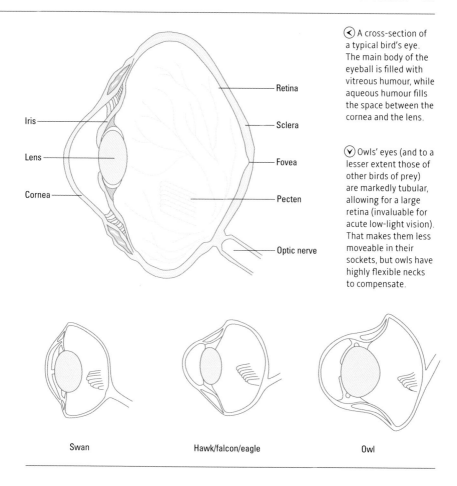

A cross-section of a typical bird's eye. The main body of the eyeball is filled with vitreous humour, while aqueous humour fills the space between the cornea and the lens.

Owls' eyes (and to a lesser extent those of other birds of prey) are markedly tubular, allowing for a large retina (invaluable for acute low-light vision). That makes them less moveable in their sockets, but owls have highly flexible necks to compensate.

Iris

Lens

Cornea

Retina

Sclera

Fovea

Pecten

Optic nerve

Swan

Hawk/falcon/eagle

Owl

to a minimum. The pecten is pigmented to help protect the blood vessels it contains from ultraviolet light.

A bird's retina contains rod cells, which respond to changes in light and shade, and cone cells, which are sensitive to light of particular wavelengths (colours). Avian cone cells have four or possibly five kinds of photopigments, and some cones are double. Human eyes have only three types of cones. Birds are therefore sensitive to a wider range of wavelengths than us, giving them a visual

field that extends into the ultraviolet spectrum. The fovea is the point on the retina where cone cells are especially highly concentrated, giving the sharpest vision. Some birds have two or even three distinct foveae.

NOCTURNAL BIRDS

The overwhelming majority of bird species are diurnal – most active in daylight hours. Others are crepuscular – most active in the twilight periods of dawn and dusk. Those birds that are most active at night have particular adaptations to help them live in darkness.

The most familiar of all nighttime birds are the owls, though many species are not strictly nocturnal. The nighthawks and nightjars, and their allies the potoos, frogmouths and the Oilbird (*Steatornis caripensis*), are also mostly nocturnal. Other nocturnal birds include the kiwis, the thick-knees or stone curlews, some members of the rail family, Jerdon's Courser (*Rhinoptilus bitorquatus*), an Indian species believed to be extinct until its rediscovery in 1986), and the Swallow-tailed Gull (*Creagrus furcatus*) of the Galápagos Islands.

One of the challenges nocturnal birds face is how to hide during their daytime sleep, when other birds are active. A sleeping bird in the open is easy prey, and even the largest owls may be driven from their roosts by other

⊼ The retinas of nocturnal owls' eyes contain a high proportion of rod cells, which are acutely sensitive to light contrast, though not to colour.

birds. For this reason, many nocturnal birds have superbly camouflaged plumage, and a body shape that helps conceal them. The ear tufts of some owls disguise their silhouette when they are perched, making them look like broken branches. Nightjars have a hunched, low profile and sleep perched along rather than across branches.

Nocturnal birds' eyes are large to gather more light, and have far more rod cells than cone cells in the retina. This means they are less sensitive to different colours, but more so to changes in contrast. Many also have a reflective membrane (tapetum lucidum) behind the retina, to gather extra light. Hearing becomes more important in darkness – owls are noted for their exceptionally acute and accurate hearing. Male owls, though smaller than females, have larger syrinxes (see page 113) to help them produce their resonant, far-carrying territorial calls. However, owls out hunting need to be silent so they can hear prey without the prey hearing them. Their adaptations for this include modified flight feathers, with rough comb-like leading edges to break up airflow and quieten the sounds of their wingbeats.

Seabirds at night

Some species of petrels and shearwaters spend the daylight hours at sea, and only come to land at night. Equipped with a fine sense of smell to locate food while out at sea, a storm petrel uses this same sense on land to home in on its nesting burrow at night. For the Swallow-tailed Gull, the opposite is true – it spends the night out at sea foraging and comes to its nesting colonies in the daytime. It has exceptionally large eyes with a high light-gathering capacity, to help it see its prey – fish and squid that come to the sea surface at night.

Night migration

Many small songbirds migrate at night, a safer time to be on the wing when fewer predators are active. They are guided by geomagnetic information, and probably also by star positions, and travel in flocks that follow the leading birds and stay in contact through constant calling. However, night migration is curtailed on the darkest, most overcast nights.

⏚ The Swallow-tailed Gull is the only gull species that feeds mainly at night. Its large eyes are an adaptation to low-light foraging.

VARIATIONS IN AVIAN EYES

Predators need to focus on their target, while prey species need to spot danger approaching. Nocturnal birds need eyes that work well in low light, and aquatic birds need adaptations to allow them to see through water. These requirements mean differences in the eyes' appearance and anatomy.

The position of the eyes on a bird's head determines its field of view or visual field. Birds with front-facing eyes have limited side-on vision, but their eyes' fields of view overlap extensively, which allows for better judgement of distance. This is typical of predators, most obviously the owls, which need a clear view of their prey and a perfect understanding of the distance they need to cover to reach it. For prey species, a wide field of view is more important, so these species have eyes set farther to the sides of their head. Woodcocks' eyes are particularly high and far back on their heads, giving them almost a full 360-degree visual field.

Nocturnal owls' eyes have relatively few colour-sensing cone cells in their retinas, but a high concentration of rods, to detect the shifts in brightness that reveal the smallest movements. Some nocturnal and crepuscular birds also have a tapetum lucidum – a shiny membrane behind the retina that bounces

the light back through, giving the retinal cells a second opportunity to react to it.

Birds that find their prey near the surface of water have droplets of red oil in their retinal cone cells. This helps them to overcome the distortion of vision that results from looking through the air into the water. Birds that hunt their prey while swimming underwater have particularly flexible lenses in their eyes to quickly make the adjustment from in-air vision to underwater vision. This adaptation also compensates for their flatter corneas (to protect them from the effects of water pressure).

EYE COLOUR

From family to family and species to species, birds exhibit a great variety of eye colours, from almost black to almost white via all shades of yellow, orange, brown, red, blue and green. In many species, eye colour changes with age. Young Herring Gulls (*Larus argentatus*) are dark-eyed but by adulthood their eyes are very pale; in several raptors in the genus Buteo the reverse occurs, eyes darkening with age. Eye colour seems to have no definite significance with regard to visual ability, but blinking to 'flash' brightly coloured eyes is part of the courtship display in some species.

(‹) Gulls are predators as well as scavengers and have a fairly forwards-oriented field of view.

ⓐ Some examples of the varied eye colours found in the bird world. Clockwise from top left:
Tree Sparrow (*Passer montanus*); Eagle Owl (*Bubo bubo*); Jackdaw (*Corvus monedula*); Cormorant
(*Phalacrocorax carbo*); Hen Harrier (*Circus cyaneus*); Victoria Crowned Pigeon (*Goura victoria*).

EARS AND HEARING

Although birds lack the outer ear structures seen in most mammals, you only need to listen to the exuberant and varied birdsong in a springtime forest to understand how important sound is in the lives of birds.

If you look at a young songbird that has yet to grow its first feathers, you will easily spot the large ear openings on the side of the head, each one set behind and just below the eye. In mature birds, the ear covert feathers lie across the openings, concealing them, though they can be seen in bareheaded birds like turkeys. The feather tufts that distinguish some species of owl are to disguise their outline and have no relationship to hearing, though the facial disks of owls and harriers do help channel sound into their ear openings, and other birds can do the same by raising their ear covert feathers. Birds can

also contract a muscle in the surrounding skin to close off the ear opening.

Sound waves passing into the ear will cause the eardrum, a conical membrane across the ear canal, to vibrate. These vibrations are passed on to the middle ear, where they are greatly amplified. The vibrations are then transmitted to the cochlea in the inner ear, where sound waves are converted to nerve impulses. The cochlea is a fluid-filled small bony tube, lined on its inner surface with sensory hair cells. Vibrations passing through the fluid in the cochlea move the hair cells. When they are thus stimulated, they transmit neural impulses, which pass via the vestibulocochlear nerve to the sound-processing regions of the brain.

ⓥ Through sound, a territorial male House Sparrow (*Passer domesticus*) can advertise his presence to others without having to be in plain view.

ⓡ The prominent ruff of stiff feathers around a Barn Owl's (*Tyto alba*) face helps it channel sounds into the ear openings on the sides of its head.

The inner ear also contains the apparatus for the vestibular (balance and motion) sensory system.

AUDITORY ACUITY

Sound is important for more than just communication. Birds use their hearing to sense danger, to find prey (for example, thrushes can hear worms moving in soil, and woodpeckers can hear beetle larvae gnawing deadwood inside tree stumps), and for navigation.

Most birds have good hearing at a range of frequencies similar to that audible to humans. However, their acuity is better, and in the case of owls it is extraordinarily refined. The most strictly nocturnal owls can locate their prey by sound in near total

darkness, or attack it through a concealing layer of snow. Many owls have asymmetrical ear openings (and, in some cases, skulls – see pages 34–35) so that sounds directly below and above the bird reach the ears at slightly different points, allowing the owl to pinpoint any sound in three-dimensional space. Oilbirds, which nest in dark caves, use simple echolocation to find their way around – like bats, they make clicking calls then listen for the echoes deflected from solid surfaces.

SMELL AND TASTE

For most birds, the senses of smell and taste are very much secondary to vision and hearing. However, taste and smell are still important to them in many ways.

It was once supposed that birds in general had virtually no sense of smell at all, but we have now learned that they do perceive and respond to odours – not just in finding food, but in communication and navigation. Birds' nostrils are a pair of holes on the upper mandible of the bill, usually closer to the base than the tip. The bird senses odours via cells in the olfactory epithelium. This membrane lines the nasal conchae or turbinates – complex swirls of bone within the nostrils that warm up inhaled air. Receptor cells (modified neurons) within the olfactory epithelium extend cilia – fine threads – into the space within the conchae, and these cilia capture odour molecules. When they do so,

they send nerve impulses along their fibres. Ultimately, the individual fibres come together to form the olfactory nerve, which carries information to the olfactory bulb in the brain.

There is great variation in the size of the conchae and of the olfactory bulb. The largest can be found in the 'tubenose' seabirds (albatrosses, petrels and shearwaters), and the American vultures. These carrion-consuming birds can detect the scent of a carcass from up to 12 miles (20km) away. New Zealand's flightless kiwis, which forage nocturnally on foot in undergrowth and probe damp soil, are also capable smellers, with their nostrils positioned at the tips of their long bills. The olfactory bulb is just 10 per cent of the length of the whole brain

ⓥ Some seabirds have an acute sense of smell, helping them to locate floating food over long distances.

in most songbirds, but in the Turkey Vulture it is 29 per cent, in kiwis 34 per cent, and in some petrels almost 40 per cent.

(∧) The scent of carrion can attract Turkey and Black Vultures (*Cathartes aura* and *Coragyps atratus*) from a distance of many miles.

TASTE BUD COMPARISON

Birds have limited tasting abilities compared to mammals, and their anatomy for this sense is rather different too. Mammals have taste buds in their tongues, which contain receptor cells that respond to sweet, salty, bitter, sour and umami (savoury) flavours. A human tongue contains about 10,000 taste buds, but the tongue of a bird has very few. Further study has revealed that avian taste buds are mainly located elsewhere – in the membranes lining the inside of the mouth, rather than in the tongue. Nevertheless, the number of taste buds is much lower than in mammals (just 400 or so in the Mallard, *Anas platyrhynchos*), although the range of taste appears to be similar, with receptor cells detecting sweet, salty, bitter and sour flavours

at least. The receptor cells work in a similar way to those in the olfactory epithelium, picking up molecules through cilia and conveying the information via nervous impulses to the brain.

Birds with the most well-developed sense of taste include nectar-feeding hummingbirds, which have been shown to prefer more concentrated sugar solutions, and fruit-eaters like parrots, which will reject sour-tasting unripe fruits. Some shorebirds, which find prey by probing into mud or sand, can sense by taste whether burrowing worms are nearby.

SKIN ANATOMY AND TOUCH

Sensing through touch is essential to birds, particularly when it comes to foraging and to close communication with others of its species. The bill is especially sensitive to touch and enables some birds to create a detailed mental map of the immediate surroundings.

Birds have four kinds of receptor nerve cells in their bodies for tactile sensation, in the form of pressure, vibration and stretching. These are Herbst corpuscles, Grandry corpuscles, Merkel receptors and Ruffini endings. The Herbst corpuscle is the most abundant type, though Grandry corpuscles are abundant in the bills of aquatic birds. Both corpuscles have layered, oval capsules of collagen surrounding the nerve ending, and any distortion to the capsule is conveyed through the layers of the capsule to the nerve ending, triggering it to send a

ⓥ A well-developed sense of touch enables macaws to use their massively powerful bills as gentle grooming tools.

nerve impulse. The other two cell types are not encapsulated but are free nerve endings. In the skin, sensations are often conveyed via movement of feathers – the part of the feather shaft that is embedded in the skin pushes against nearby tactile receptors in the skin.

Many birds use their bills to probe, explore and dabble, using touch rather than other senses to find their food. The kiwis, nocturnal and with very poor eyesight, use a combination of smell and touch to find invertebrates in soft ground. Shorebirds including curlews and sandpipers have good eyesight but when they are feeding they are foraging 'blind'. The action of a shorebird's bill probing into wet sand sends a pressure

Ⓐ Waders can sense the movement of nearby burrowing worms or other prey when they probe soft ground with their bills.

Ⓐ Owls and many other birds have highly touch-sensitive filoplumes around their bills, used in close-range interactions such as mutual preening.

wave through the surrounding water in which the sand grains are suspended – a wave that is blocked if it meets a solid object, such as a burrowing mollusc. By probing repeatedly and rapidly ('stitching') and sensing the pattern of vibrations from the previous pressure waves, the bird can quickly work out the positions of possible food items.

Ducks' bills have a concentration of pits at the bill tip that contain Herbst and Grandry corpuscles and a few Ruffini endings – you can see the tiny depressions if you look closely. The collection forms what is known as the bill-tip organ, and the information it gathers, along with data from the taste buds, help the duck determine whether an item it picks up while dabbling is food or not.

FEATHERS AND TOUCH

Sensing touch to the skin guides a bird's preening behaviours, which is crucial for keeping its feathers ordered, in good condition, and free of debris. Birds cannot effectively preen all of their own head plumage (or neck plumage in the longer-billed species) but they will preen each other's – this particular form of touch helps to build and cement bonds between a breeding pair, or the members of a family group.

Some specialised feathers are effectively organs of touch. Barbless, hair-like filoplumes are present over the whole body. Those in the wing are connected by nerves to the cerebellum – the brain structure involved in coordinating movements – their input helps the bird adjust its position as required during flight. Filoplumes around the face can help the bird handle its food and feed its chicks with the correct degree of care – this is important to owls, for example, which have poor close-range vision. The sensitivity of facial filoplumes enables birds to close their eyes for protection when using their bills to deal with potentially dangerous prey.

EXTRA SENSES

Beyond sight, hearing, smell, taste and touch, birds (and other animals) have additional senses to help them draw on information about the outside world. The most remarkable and alien (to us) of these is their ability to sense magnetic fields.

Earth's magnetic field is created by its moulten iron interior, which acts as a giant magnet with poles roughly aligned to the geographic North and South Poles, and magnetic field lines looping around Earth between the two poles. This field, the magnetosphere, helps to protect us from solar radiation, and it also provides a means of navigation for those animals that can detect it.

Birds possess crystals of magnetite (magnetic iron oxide) in front of their eyes and deep inside their nostrils, which gives them the means to sense Earth's magnetic field. In some species, it has also been demonstrated that light entering the eyes produces a different retinal response when the bird is in the presence of a magnetic field, meaning the eye can act as a compass. We do

not yet fully understand the concept of avian magnetoception but it could be that birds can effectively 'see' magnetic field lines – there is strong evidence that magnetic fields alone are enough to change their navigational behaviour when migrating. Interestingly, in

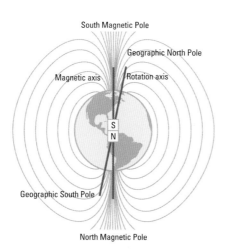

South Magnetic Pole

Geographic North Pole

Magnetic axis

Rotation axis

S
N

Geographic South Pole

North Magnetic Pole

(‹) Earth's magnetic field, generated by its core of moulten iron, is a reliable aid to navigation and is used by many migratory birds and other animals.

④ Birds of prey often manoeuvre their wings and body dramatically in the air – the vestibular system helps them to keep their heads steady.

④ An ability to sense Earth's magnetic field helps birds like the Robin (which is migratory in northern parts of its range) to navigate correctly during migration.

the Robin and a few other species it has been found that the 'eye compass' is operational in the right eye only in adulthood, the left eye compass being lost in the first year of life.

A bird also possesses receptor nerve endings for sensing changes in temperature and pain. Both thermoceptors and nociceptors are free nerve endings in the skin. The pain-sensing nociceptors are modified forms of touch- and heat-sensing cells with a high threshold – they are only stimulated when pressure or a temperature change is strong enough to cause injury.

IN BALANCE

The vestibulocochlear nerve, which carries auditory information to the brain, is also concerned with transmitting signals from the vestibular system, concerned with balance. The vestibular system is located in the inner ear, above the cochlea, and consists of three fluid-filled loops, the semicircular canals, each oriented in a different direction and all connected to a larger space called a vestibule. The vestibule contains two sacs of sensory cells, called the otolith organs. The vestibular system's function is to send information to the brain about where the bird's head is in space and in relation to its body, so that it senses tilting and rotational movement. The

hair cells in the otolith organs transmit information about the movement of fluids in the vestibule to the brain, via the vestibulocochlear nerve. The vestibular system helps a bird to keep its head level while it tilts and turns its body through rapid aerial manoeuvres. You may notice, when you watch a bird such as a Fulmar (*Fulmarus glacialis*) or a kite in circling flight, that even when it tilts so far that its wings are held on a vertical line, its head remains absolutely level.

THE CIRCULATORY SYSTEM

Blood is the method by which oxygen – vital for all energy-consuming physiological activity – reaches every part of the body. Birds often expend vast amounts of energy in their daily lives and therefore need to keep their blood pumping strongly.

> The muscles of an active hummingbird consume oxygen ten times faster than those of a top-level human athlete working at maximum effort.

THE BIRD CIRCULATORY SYSTEM

All the living tissues in a bird's body carry out various tasks. To achieve this, they need to be supplied with energy and their waste products to be taken away. The circulatory system is responsible for this delivery and disposal.

In birds, with their extremely high-energy lifestyles, the circulatory system is very hard working indeed. The bloodstream provides transit of oxygen, nutrients, hormones, waste products and other metabolites to the muscles, organs and other body tissues as required. Blood is carried around the body via a network of blood vessels, and kept moving by the pumping of the heart. The arteries are outbound blood vessels, carrying oxygenated blood to the body tissues, while de-oxygenated blood returning towards the heart is carried along veins.

⊙ The heart rate of a hummingbird may quadruple as soon as it takes off from its perch and begins to hover and fly.

CIRCULATORY SYSTEM COMPARISON

The circulatory system of birds is similar to that of mammals. The heart is four-chambered, and blood flows along a double-loop system, with de-oxygenated blood pumped to the lungs (the pulmonary circuit) and then, once oxygenated, returning to the heart to be pumped to the rest of the body (the systemic circuit). Birds' hearts, however, are larger, beat more rapidly, and pump a larger volume of blood than in mammals of the same size. In general, the smaller the bird, the quicker its heart rate, and the smallest and most frantically active of all birds – the hummingbirds – have the fastest heart rates. A Blue-throated Hummingbird's (*Lampornis clemenciae*) heart rate can reach 1,260 beats a minute when the bird is active. The hummingbird heart rate is still rapid at rest – some 250 beats a minute – but can drop down to less than 100 when the bird is in a torpid, energy-conserving state on a cold night.

The body also has a system of organs and vessels for lymphatic circulation. This system is concerned with removing excess fluids and proteins from tissues and returning them to the blood, and is also a component of the immune system, carrying white blood cells that recognise and destroy bacterial and virus-carrying cells within the body.

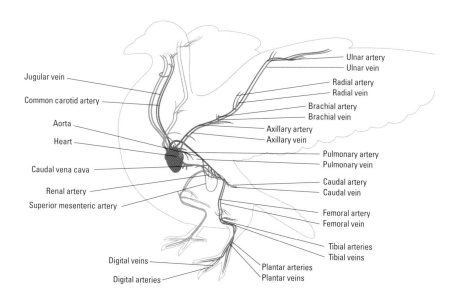

Jugular vein

Common carotid artery

Aorta

Heart

Caudal vena cava

Renal artery

Superior mesenteric artery

Digital veins

Digital arteries

Ulnar artery
Ulnar vein

Radial artery
Radial vein

Brachial artery
Brachial vein

Axillary artery
Axillary vein

Pulmonary artery
Pulmonary vein

Caudal artery
Caudal vein

Femoral artery
Femoral vein

Tibial arteries
Tibial veins

Plantar arteries
Plantar veins

ⓥ Like other ratites, the Emu has a relatively slow heart rate, comparable to a human's.

ⓐ The main vessels in the blood circulatory system of a pigeon.

THE BIRD HEART

A perfectly functioning heart is crucial to survival for wild birds. This muscular organ is simpler in its general structure than most other vital organs, but on the cellular level, it is a finely tuned machine, its activity regulated by a series of precisely coordinated electro-chemical processes.

The avian heart consists of four chambers – a left and a right atrium on top, and a left and a right ventricle below – all contained in a thick membrane called the pericardium. There is no communication directly between the left and right sides, but the atrium/ventricle pairs on each side are connected via valves. Freshly oxygenated blood from the lungs enters the heart via the left atrium, and passes to the left ventricle. De-oxygenated blood that is returning to the heart enters via the right atrium, and then is pumped to the lungs via the right ventricle. Because the left ventricle has to pump the blood much farther than the right, it is much larger, with more muscular walls.

The cardiac muscle that forms the heart is similar to skeletal muscle in structure but its contractions are involuntary. In the right atrium, there is a cluster of specialised pacemaker cells (p-cells), which together form the sinoatrial (SA) node. These cells, in their resting state, have a negative electrical charge. When nerve impulses pass into the SA node, it causes their charge to switch from negative to positive. This state change is passed on to nearby cardiac muscle cells and spreads rapidly, causing the two atria to contract. The electrical impulse passes downwards to the ventricles and they also contract, completing a heartbeat.

(∧) Sunbirds have a high metabolic rate and rapid heart rate, fuelled by their energy-rich nectar diet.

Should the SA node fail, a second cluster of p-cells between the left atrium and the right ventricle, the atrioventricular (AV) node, can take over the role of heart pacemaker. A third node located lower still in the heart, the bundle of His, can also fulfil the role, though both are less efficient, and their normal role is to pass on the electrical impulses that begin in the SA node.

HEART COMPARISON

Birds' hearts make up 1 per cent or more of the total body weight in most small birds, and more than 2 per cent in hummingbirds. By comparison, in humans, the heart makes up about 0.3 per cent of the body weight, in cats about 0.45 per cent and in rabbits about

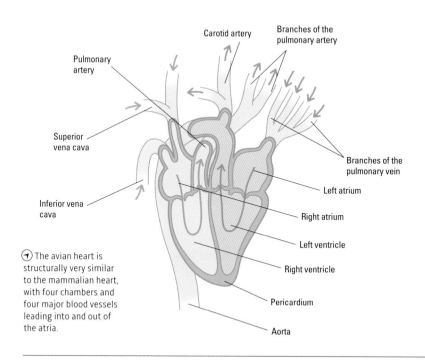

Carotid artery

Branches of the
pulmonary artery

Pulmonary
artery

Superior
vena cava

Branches of the
pulmonary vein

Left atrium

Inferior vena
cava

Right atrium

Left ventricle

⊙ The avian heart is
structurally very similar
to the mammalian heart,
with four chambers and
four major blood vessels
leading into and out of
the atria.

Right ventricle

Pericardium

Aorta

0.27 per cent. Flightless or weakly flying birds
have proportionately smaller hearts – for
example, tinamous' hearts make up about
0.2 per cent of their body weight. Some owl
species also have proportionately small
hearts, such as the Spectacled Owl (*Pulsatrix
perspicillata*), a forest species which mainly
hunts by pouncing on prey from a perch.
A bird's typical mode of flight also has a
strong bearing on heart size. Buzzards that
habitually soar and scan for carrion on the
ground have proportionately smaller hearts
than falcons, which actively chase living prey.

ⓐ The Elegant Crested Tinamou (*Eudromia elegans*) is
a chicken-like bird that is related to the ratites and has
notably small hearts for its body size.

BLOOD VESSELS

The blood vessels carry blood from the heart to the body tissues and back again, and can be thought of as rivers that produce branching tributaries and streams.

There are four major blood vessels entering and exiting the heart and the structure of vessels carrying blood away from the heart (arteries) is distinctly different to those bringing it back (veins). Freshly oxygenated blood arrives from the lungs via the pulmonary vein, into the left atrium, and exits the left ventricle via the aorta (the body's largest artery), heading for the rest of the body. On its return journey, depleted of oxygen, it enters the right atrium via the vena cava (the body's largest veins). From there it flows into the right ventricle and is pumped to the lungs via the pulmonary artery, to be replenished with oxygen. The aorta and pulmonary artery both branch into smaller arteries, while the vena cava and pulmonary vein are fed into by smaller veins. The

smallest arteries and veins are known respectively as arterioles and venules – capillaries branch directly from arterioles and feed directly into venules.

The other major veins and arteries in the body are named for the body parts they reach – branchial for those serving the wings, femoral for those to and from the legs, and so on. The hepatic portal vein carries blood from the digestive tract to the liver – it is not a true vein as it does not go to the heart.

Arteries close to the heart have thick walls full of elastic fibres, and expand to accommodate the force of blood being pumped into them. Their elastic recoil after each surge helps to push the blood along. Farther from the heart, where blood pressure is lower, the walls of arteries contain

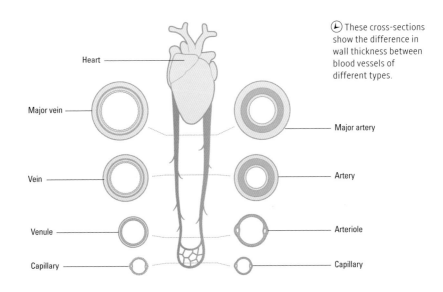

Heart

Major vein

Vein

Venule

Capillary

(L) These cross-sections show the difference in wall thickness between blood vessels of different types.

Major artery

Artery

Arteriole

Capillary

proportionately more smooth muscle and fewer elastic fibrous tissues. Veins have relatively thin walls and can distend to hold a high volume of blood. Large veins contain valves that prevent backflow of blood.

Exchange of gases, nutrients and other molecules takes place through the walls of the capillaries – the smallest vessels. Capillaries come in three types, based on how permeable their walls are. Most are continuous, and can only release and take in water molecules through tiny spaces between the cell layer that forms their walls. Fenestrated capillaries contain pores that can admit larger molecules, and sinusoid capillaries have large spaces in their membranes, big enough for whole cells to pass in and out. These capillaries are found, for example, in bone marrow, where new blood cells need to pass into the circulatory system.

⊼ A unique arrangement of blood vessels allows cold-climate birds like these Adélie Penguins (*Pygoscelis adeliae*) to walk on ice without freezing their feet.

TEMPERATURE CONTROL

As well as carrying blood, the circulatory system has a role in regulating body temperature. Arterioles and, to a lesser extent, arteries and veins can dilate or constrict to release or retain heat as needed. Birds that live in freezing conditions have a complex network of blood vessels in their feet that provide a countercurrent heat exchange system. The incoming artery, carrying warm blood, is surrounded by outgoing veins and so transfers its heat to them. This keeps the blood in the foot from freezing, and also means that the blood is warmed up as it returns to the body.

COMPOSITION OF BLOOD

The blood is the body's combined courier and disposal system, carrying nutrients, enzymes, hormones, and so on to where they are needed and removing waste products to where birds can excrete them.

Blood is primarily composed of liquid (plasma) but also contains a variety of cells for different purposes – those in birds differ in several respects to those in mammals. With the blood cells removed, blood plasma is a pale yellowish fluid, composed mainly of water (85 per cent). It carries molecules such as amino acids and proteins, glucose, electrically charged ions of sodium, calcium and other elements (electrolytes), hormones, antibodies and waste products, which pass into and out of the bloodstream through capillary walls.

The most abundant cells in the blood are the red blood cells or erythrocytes. These contain the oxygen-carrying, iron-based molecule haemoglobin, which gives blood its red colour. Oxygen molecules bond to the haemoglobin – four oxygen molecules per haemoglobin molecule – and in due course are released into the blood plasma to be taken up by the body tissues. Their release is encouraged by the presence of carbon dioxide, a waste product of cellular activity. Birds' erythrocytes are flattened, oval cells containing a cell nucleus, unlike mammals' erythrocytes, which are anucleate, having lost their nuclei during their development.

The bird erythrocyte is slightly larger than the mammal's, and there are about 2.5–4 million of them in each cubic millimetre of blood – fewer than in most mammals (for example, humans have

⌖ A hard-working circulatory system gives birds the energy they need to avoid suffering injury in an aerial fight.

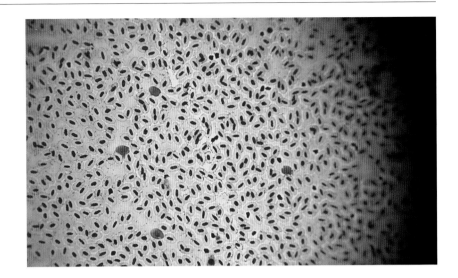

4.1–5.9 million per mm³ and cats 5.9–
9.9mm³). Bird erythrocytes survive for
between 28 and 45 days (less than half
the lifespan of those in most mammals);
as they near the end of their lives, their
membranes change in structure and they
become targeted by white blood cells
(leucocytes), which engulf them and break
them down. White blood cells are active
components of the immune system. They
include lymphocytes and a variety of types
of granulocytes (see page 108). Erythrocytes,
thrombocytes and leucocytes all develop
within the bone marrow.

CLOTTING

The blood also contains thrombocytes,
nucleated blood cells that are the avian
equivalent of the tiny, anucleate platelets in
our own blood. Like platelets, thrombocytes
have a role in blood clotting, becoming sticky
in the presence of collagen (which is present
in the outer layers of blood vessels – blood
coming into contact with collagen is

(Ⓐ) Under the microscope, the nuclei in bird red
blood cells are apparent.

therefore indicative of damage to the vessel).
Thrombocytes, though, are much larger than
platelets and they have nuclei; they are also
less sticky and do not form multi-layered
aggregations. Birds are much less likely than
mammals to develop issues with blood vessel
blockages that can lead to heart attacks and
strokes, because of the differences between
platelets and thrombocytes.

HORMONES AND GLANDS

One of the vital roles of blood is to carry hormones around the body. These protein molecules regulate a wide range of physiological processes, so the concentration of each within the blood fluctuates significantly through the day, month and year.

Hormones are released into the bloodstream from various glands and other organs, while there are also glands that secrete their contents to the outside of the body. Hormones, and the glands and organs that release them, are all part of the bird's endocrine system. The function of the various hormones is primarily regulatory and stimulatory, keeping levels of water, salts and other substances within the right range for healthy function (homeostasis), and controlling cyclical processes in the bird's body. Hormonal activity governs the reproductive cycle, sleep cycle and moult cycle, and is also involved in growth, metabolism, the balance of blood calcium and glucose levels and various aspects of behaviour. Some hormones' function is to stimulate other glands to release their hormones. The chemical structure of hormones varies – many are proteins, while others are oxidised fatty acids and steroids. They work by binding chemically to receptor sites of cell membranes in the tissues they activate, triggering various cellular processes.

The pituitary gland, located in the brain, releases seven hormones: the follicle-stimulating hormone 'ripens' ova in the ovaries; the luteinising hormone stimulates the ovaries and testes to produce the 'sex hormones'; prolactin affects the number of eggs in a clutch and the degree of parental care; growth hormone stimulates growth; adrenocorticotrophic hormone helps regulate the sleep/wake cycle; thyroid stimulating hormone regulates the activity

of the thyroid gland; and the melanotropic hormone affects deposition of melanin pigment in the body.

Other important glands include the pancreas, producing insulin and other hormones concerned with regulating blood sugar; the thyroid and parathyroid in the throat, releasing hormones that affect body temperature and food metabolism; and the adrenal glands above the kidneys, which regulate blood pressure and are involved in the 'fight-or-flight' response to stress, governed by the sympathetic and parasympathetic nervous systems. The gonads (testes and ovaries) have glandular tissue that produces the three sex hormones: oestrogen, testosterone and progesterone. All three have a role in driving sexual function and breeding behaviour.

(∧) Albatrosses never need to drink fresh water because their salt glands can extract excess salt from seawater.

NON-HORMONAL GLANDS

Externally secreting, non-hormonal glands include the uropygial gland at the base of the tail; present in most birds and largest in aquatic species, it produces preen oil, which the bird applies to its feathers to promote waterproofing, discourage parasites and sometimes to exude particular odours. The salt glands, located above the eyes in seabirds, filter excess salt from the blood and secrete it via ducts to the nostrils, where it evaporates.

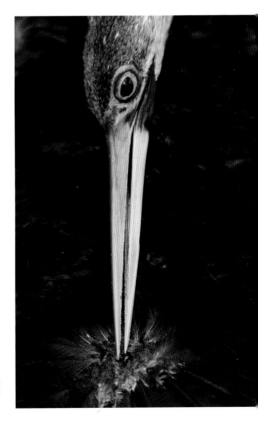

(>) An Anhinga (*Anhinga anhinga*) uses its bill to extract preen oil from its uropygial gland.

(<) Courtship behaviour, such as the throat-inflating display of male frigatebirds, is triggered by shifts in hormonal balance at certain times of year.

THE LYMPHATIC AND IMMUNE SYSTEMS

The lymphatic system is a system of vessels that carries lymph or interstitial fluid, communicates with blood vessels and has a critical role in the body's immune response.

When potentially harmful cells are present in the body, the immune system's job is to find and remove or destroy them. These cells could be bacteria or other pathogens, or cells from the bird's own body that are damaged, dying naturally, or have been 'taken over' by a virus. White blood cells or leucocytes carry out this function. Lymph vessels are not very different from blood vessels, consisting of a network of small lymph capillaries that connect to larger lymph vessels. However, because lymphatic drainage is a one-way system, rather than a continuous circuit, the capillaries are blind-ended. They have permeable walls and drain excess interstitial fluid or lymph from between cells in many kinds of tissues. The lymph is passed along larger lymph vessels, which have contractile walls and valves to prevent backflow, and eventually into the blood circulatory system via ducts in the upper chest or neck area. Some flightless birds have muscular structures called lymph hearts within the lymphatic system, which pump the lymph along the vessels (in flying birds, muscle contraction in normal activity is sufficient to squeeze the lymph along).

LYMPHOCYTES AND GRANULOCYTES

The two categories of white blood cell – lymphocytes and granulocytes – are formed in the bone marrow and carried in the lymph to the bloodstream once mature. There are two main types of lymphocytes. B-lymphocytes or B-cells produce antibodies, proteins that bind to specific structures

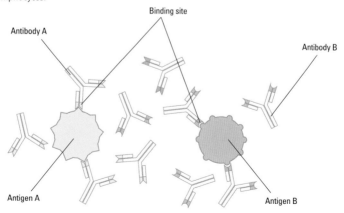

(◀) The antibodies produced by B-lymphocytes bind to antigens on the membranes of 'foreign' cells in the bloodstream, marking them as targets for the 'killer' T-lymphocytes.

Binding site

Antibody A

Antibody B

Antigen A

Antigen B

(antigens) on the membranes of pathogenic cells. The presence of antibodies marks the pathogenic cells out as targets for the T-lymphocytes or T-cells, which attack and destroy them by secreting molecules that break them down. Some kinds of T-cell are classed as 'helpers' rather than 'killers' as they help B-cells to form antibodies. B-cells and T-cells form the 'adaptive response' component of the immune system, as those that encounter a particular pathogen will proliferate, making for a quicker and more effective attack on that same pathogen in the future (an acquired immune response). The granulocytes also attack pathogenic cells, but their abundance in the bloodstream is not influenced by past activity.

T-cells mature in the thymus, a structure in the throat. B-cells mature in a lymphatic organ unique to birds known as the bursa of Fabricius, an outpouching of the bird's cloaca (where the digestive and reproductive tracts exit the body). A third kind of lymphocyte known in mammals – the natural killer (NK)

(A) When migrating birds, such as these Snow Geese and Pintails (*Anas acuta*), gather in close proximity, diseases can easily spread among them.

cell, which attacks virus-infected cells and also tumor-forming cells – is probably also present in birds.

Granulocytes are not as abundant as lymphocytes in birds but are more motile (able to move around through the bloodstream and to enter tissues) and so are particularly quick to respond to any abnormal or alien cells. They are phagocytes, destroying the target cell by engulfing it within their membrane and lysing (breaking down) its contents.

THE RESPIRATORY SYSTEM

The breathing system of birds is an elegant marvel of nature, combining efficiency and close integration with other bodily systems to enable some birds to sing continuously, fly hard for long periods and even do both of these things at the same time.

⊙ Singing birds can produce sound at the same volume at all points in their breathing cycle.

THE BIRD RESPIRATORY SYSTEM

Birds' bodies require a great deal of oxygen, particularly when in flight. Oxygen is crucial to releasing energy, but the process generates carbon dioxide, which is harmful in excess and must be removed from the bloodstream. Birds have a complex respiratory system to meet both needs.

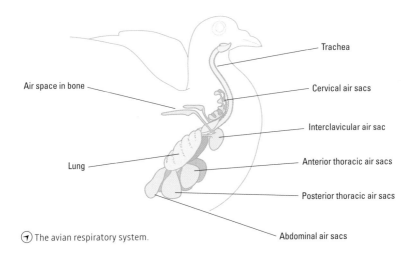

Air space in bone

Lung

Trachea

Cervical air sacs

Interclavicular air sac

Anterior thoracic air sacs

Posterior thoracic air sacs

Abdominal air sacs

⊙ The avian respiratory system.

Like mammals, birds inhale air through their nostrils or occasionally their open mouths, and the air passes down the windpipe or trachea on its way to the lungs. Within the lungs, gaseous exchange takes place, with oxygen passing into the bloodstream via capillaries (see pages 102–3), and carbon dioxide diffusing from the capillaries into the air spaces in the lungs (see pages 114–15).

Birds' lungs are proportionately small compared to those of mammals, and they are also rigid, rather than expanding and contracting with each inhalation and exhalation. The lungs are only part of the story when it comes to the breathing cycle in birds, though. A far larger part of the internal body space is occupied by a system of air

sacs, which draws air through the lungs. Because of the way the air sacs are arranged, the breathing cycle of birds is unidirectional, air flowing only one way. It is therefore much more efficient than that of mammals, with each breath drawing completely fresh air through the lungs, rather than fresh air mixing with a residual amount of 'old air'.

This breathing system allows birds to be fully active in conditions with much lower oxygen than a human could tolerate. The Bar-headed Goose (*Anser indicus*), for example, famously crosses over the Himalayas on its annual migration and has even been seen flying over Mount Everest. This feat entails flying at heights of more than 7,000 metres (23,000 feet), where oxygen levels are just 10 per cent of that of air

at sea level. The high-altitude air is also much less dense, making the goose's fast-flapping flight even more energy-demanding.

⊗ Bar-headed Geese have been recorded flying at extremely high altitudes, where the oxygen levels are extremely low, without apparent difficulty.

BREATHING AND SINGING

The unusual and efficient avian breathing system also accounts for their vocal abilities, in particular the ability to sing continuously while inhaling and exhaling. The syrinx, a structure at the base of the trachea, is the organ of sound production (see pages 118–19), but the trachea itself is usually a relatively simple tube of tightly stacked cartilaginous rings. It opens in the base of the mouth, via a valve called the larynx. Although this structure is known as the 'voice-box' in humans, in birds it is not involved in sound production.

⊗ For their size, small birds such as the Bluethroat (*Luscinia svecica*) produce loud song that carries some distance.

LUNG ANATOMY

The sensation of taking a deep breath must be very different for a bird than it is for us. Our lungs act as bellows, expanding and pulling air in as we breathe, but in birds, the bellows are the air sacs, and the lungs are inflexible.

The efficient avian breathing system allows birds' lungs to be relatively small – only half the size of the lungs of similar-sized mammals – and their microstructure is also somewhat different from that of the mammalian lung. The airflow pattern is such that a complete breathing cycle, to move the same breath of air through the entire system, comprises two sets of inhalations and exhalations. Part two of each breathing cycle therefore occurs at the same time as part one of the next cycle.

The stiff tube of the trachea divides into two branches (primary bronchi) just after the start of the syrinx, and each primary bronchus enters one of the two lungs and exits at the other end, into the posterior air sac system (see pages 116–17). Each primary bronchus has side branches (secondary bronchi) along its length, but on the first inhalation the air mostly bypasses these, being drawn along the bronchus directly through to the posterior air sac system in the rear half of the body. On the first exhalation, abdominal muscles squeeze these air sacs, pushing the air back into the bronchus and back into the lungs, for gas exchange.

The secondary bronchi divide into several hundred air capillaries, or parabronchi. These, like blood capillaries but unlike the alveoli within mammalian lungs, are continuous rather than 'dead-ends'. The network of air capillaries is in close contact with a similarly dense network of blood

ⓥ Crocodilians and birds show considerable similarities in their breathing anatomy – a legacy of their close evolutionary relationship.

(∧) At their lek, Black Grouse (*Tetrao tetrix*) call constantly. This is facilitated by the unique avian breathing system.

capillaries, and gas exchange takes place through their walls. The oxygen-depleted air leaves the lungs via the bronchi, and moves to the anterior air sacs in the front half of the body on the second inhalation. From there, it leaves the body directly, via the trachea, on the second exhalation, without moving through the lungs again. This arrangement means that only oxygen-rich, newly inhaled air ever passes through the lungs.

RESPIRATORY SYSTEM COMPARISON

Mammalian lungs are quite different to avian lungs, being elastic and expandable, and also in having closed-ended, grape-like alveoli rather than continuous air capillaries as the gas exchange point. Without air sacs, all air that goes into the lungs returns via the same path. The lungs sit within a fluid-filled space (pleural cavity), which allows them to expand – the diaphragm (a muscular sheet below the lungs) curves downwards when they do. Birds lack both a pleural cavity and a diaphragm.

Crocodilians, the closest living relatives to birds, also have a unidirectional airflow through the lungs, but they lack air sacs; instead, they have double bronchi to each lung, and air flows into one and out via the other. There is fossil evidence that many dinosaurs had a respiratory system similar to that of modern birds.

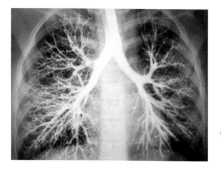

(<) Human lungs and those of other mammals are proportionately much larger than birds' lungs and also much less rigid.

THE AIR SACS

A bird's system of air sacs can transport a substantial volume of air through its lungs with high efficiency, and also hold and store air if needed.

Between them, the bird's air sacs (nine in most species) occupy a large amount of space in its body (around 15 per cent by volume), and can expand to accommodate more than ten times the amount of air that would fill the lungs. The air sacs form two distinct groups: the posterior group sits behind the lungs and comprises a pair of abdominal air sacs and a pair of posterior thoracic air sacs, while the anterior group, level with or in front of the lungs, comprises a pair of anterior thoracic air sacs and a pair of cervical air sacs. The ninth is the single interclavicular air sac, which also has extensions reaching into the air spaces in the humerus bones (see page 112).

(⌄) The air sacs provide a reserve of oxygen that helps diving birds like the Cormorant (*Phalacrocorax carbo*) to stay underwater for longer.

The air sacs contain no lung tissue and play no role in gas exchange – their role is simply to hold and move air around the body. Their walls are thin and without a blood supply, formed from very stretchy, translucent connective tissue. They do not contain muscle tissue; instead, their expansion or contraction is driven by contractions of the abdominal and thoracic muscles around them. This action means that the sternum moves forwards and backwards, as well as the rib cage opening up and contracting, through the breathing cycle, and for this reason it is very important never to squeeze on a bird's chest when you hold it. Preventing the sternum from moving will quickly cause suffocation.

Occasionally, air sacs can rupture, typically as the result of trauma. Birds that suffer this injury develop an obvious swelling under the skin from the uncontained air.

> The Kingfisher's (*Alcedo atthis*) hunting dives are brief but very costly in terms of energy, so extra oxygen stored in the air sacs may be vital.

COORDINATION IN ACTION

The breathing cycle and thus inflation and deflation of the air sacs works in concert with the wingbeats when a bird is in flight, with its chest muscles contracting and relaxing to bring the wings down and up again. There is also coordination with the leg-stepping cycle for birds that are running, or swimming on the surface, and with the wingbeat cycle for birds that use their wings to propel them when swimming underwater.

For diving birds, air held in the air sacs is utilised as an oxygen store while underwater, and in very deep-diving species, such as penguins, it probably also helps protect the lungs and trachea from being crushed by water pressure at depths. Penguins have large air sacs but have more solid bones than flying birds so do not have the additional air capacity that pneumatised bones provide.

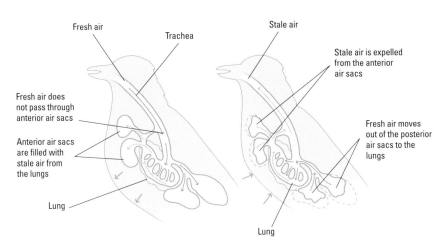

Fresh air

Trachea

Stale air

Stale air is expelled from the anterior air sacs

Fresh air does not pass through anterior air sacs

Anterior air sacs are filled with stale air from the lungs

Fresh air moves out of the posterior air sacs to the lungs

Lung

Lung

↷ Airflow within the avian respiratory system.

VOCAL APPARATUS

Birds are extraordinarily accomplished when it comes to vocal sound. Many produce beautiful natural song that has inspired writers and musicians for centuries, while others are capable of flawless mimicry of all manner of sounds, from human speech to industrial machinery.

Even among the less accomplished vocalists, each species has its own particular breath-generated sound, whether trumpeting, piping, hooting, belching, groaning, warbling, grunting, whistling, or wailing, allowing the birds to recognise one another and communicate, and birdwatchers to identify each species they hear. The bird's sound-generating organ is the syrinx, the avian voicebox. It is an organ unique to birds, and its structure, along with the unique avian breathing cycle, is the reason birds are so vocally accomplished.

Part of the respiratory tract, the syrinx is (in almost all birds) made up of the very base of the trachea, and also the uppermost parts of the two primary bronchi. Its outer wall is formed from cartilage, like the main part of the trachea, and it is surrounded by the syringeal muscles, which can change its shape by contracting. This affects the pitch of sounds that the bird makes.

At the junction where the bronchi branch apart, their walls change from cartilaginous to membranous for a short stretch. These folded tympaniform membranes generate sound when vibrated by exhaled air, as does the pessulus, a slim bar of cartilage at the centre of the junction. In a small number of birds (including owls and nightjars), the syrinx only occupies the trachea and not the bronchi, while in a few others (antbirds and their relatives, and the Oilbird) there are two separate syrinxes in each bronchus, not reaching the trachea.

MULTIPLE SOUNDS

Because the syrinx usually includes the top parts of both bronchi, it is possible for two different sounds to be made simultaneously, one by each branch. Sound can also be made continuously through repeated breathing cycles, permitting the non-stop songs of birds such as nightjars and certain warblers. Oilbirds, navigating in the darkness of their roosting caves, constantly produce sounds in

(↑) Mynas (*Acridotheres tristis*) are among the birds famed for their skilled mimicry of sound – a feat made possible by the avian syrinx.

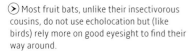 The Large-tailed Nightjar (*Caprimulgus macrurus*), like its relatives, has a tracheal syrinx, which it uses to produce its frog-like calls and song.

their two separate syrinxes that sound like single clicks to our ears but are composed of a wide range of frequencies. By listening for the echoes from their calls, they can sense the whereabouts of other birds and the cave walls, making them one of only a handful of bird species that can echolocate. However, echolocation as sophisticated as that seen in bats has never evolved in birds. In most dark, forested environments, bats dominate as nocturnal aerial insect-hunters, but the nightjars fare better in open habitats where they can use sight, rather than sound, to find and track their prey.

Most fruit bats, unlike their insectivorous cousins, do not use echolocation but (like birds) rely more on good eyesight to find their way around.

119

BIRD SONGS AND SOUNDS

Birds are charming enough in appearance and character, but delight us even more with the beauty of their voices. Their unique and impressively efficient air circulation system makes them capable of a tremendous range of sounds.

While some species have a very limited vocal repertoire, many others are absolute virtuosos. Birds of the order Passeriformes, the largest of all bird orders, are also colloquially known as songbirds. Among them are well-known songsters such as the Nightingale (*Luscinia megarhynchos*) and the Skylark (*Alauda arvensis*) in Europe, both of which have been immortalised in poetry praising the beauty of their voices. In the Americas, the Wood Thrush (*Hylocichla mustelina*) and Musician Wren (*Cyphorhinus arada*) are much admired. Songbirds tend to produce sweet and tuneful songs, although the intent behind them is more prosaic – the song is the male's way (occasionally the female's too) of discouraging rivals from entering his territory, as well as an advertisement to potential mates.

Besides the songbirds, other well-known avian vocalists include the nightjars and nighthawks with their ethereal, haunting churrs, the cranes with their loud bugling notes, the resonant and sweet-toned rippling calls of waders and shorebirds, and the bass-note booms of bitterns and eagle owls. All of these voices come courtesy of the remarkable syrinx, the organ at the bottom of the bird's trachea, with its tympanform membranes that resonate on both the inhale and the exhale. This is how a Grasshopper Warbler (*Locustella naevia*) can give its fishing-reel song non-stop for minutes on end, and how a Skylark can sing without pause while flying hard to escape a predator – the constant stream of song is a signal of

⌃ Grasshopper Warblers produce a continuous, reeling song of very rapid clicking sounds, resembling the sound of a grasshopper.

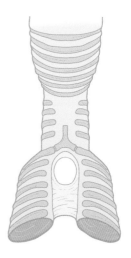

⌃ A typical songbird syrinx.

Ⓐ Nightingales, like many accomplished avian songsters, are shy by nature and have drab plumage.

its fitness, discouraging the predator from continuing its pursuit.

Alternative sounds

Some birds make their territorial sounds through other means. Snipes produce theirs by fanning out their tail feathers in fast, diving flight. The outer tail feathers are much less rigid than the inner ones, and flap back and forth in the air when the bird flies fast, creating a strange, bleating note. Woodpeckers 'drum' very rapidly on hollow, resonant wood, producing a short phrase of hammered 'notes'. Their hugely long tongues, which wrap around the back of the brain, help shield them from head trauma while they drum. Several bird species, including the Common Woodpigeon (*Columba palumbus*), some nightjars and the Short-eared Owl (*Asio flammeus*), produce loud wing-claps in

flight, as part of their courtship 'song'. This sound is produced simply by allowing the wings to strike each other above the body at the end of the upwards stroke.

As we saw in Chapter 7, some birds have modified tracheae and syrinxes that enable them to produce louder sound. The Kakapo, a curious flightless parrot native to New Zealand, has a different way of increasing the effectiveness of its calls. The male digs several saucer-shaped 'bowls' in the ground, and these serve to amplify his deep, booming calls, allowing them to carry for about 3 miles (5km). Females that are impressed by the sound will walk across the forest floor to find the male and mate.

GAS EXCHANGE

The cells in a bird's body constantly consume oxygen and release carbon dioxide. The interplay between these two gases is a crucial part of the respiratory system.

When life began to evolve on Earth, our planet's atmosphere was rich in carbon dioxide and water vapor. The earliest green plants – marine algae – made use of these two gases to generate glucose, a fuel source for their metabolic reactions, and released oxygen as a by-product. This chemical reaction, catalysed by energy from sunlight, is photosynthesis, and as photosynthesising plants spread worldwide, their activity raised the level of oxygen in Earth's atmosphere. Animals evolved to use this oxygen in their own cell chemistry, in a reaction that released energy from their stored glucose. This reaction, aerobic respiration, generates carbon dioxide as a by-product – it is exhaled and thus returned to the atmosphere.

The balance of carbon dioxide and oxygen in our atmosphere keeps plants and animals alive, and dependent on one another. In atmospheric air, molecules of the various gases (primarily nitrogen, but 21 per cent oxygen, 0.04 per cent carbon dioxide and 0.4–1 per cent water vapor) are distributed evenly, because gases move from areas of high concentration to areas of low concentration (diffusion). This process occurs with gases dissolved in liquids, as well as gases moving through other gases.

When oxygen moves through the membranes of the air capillaries of the lungs and passes through the membranes of the surrounding blood capillaries, it rapidly combines with haemoglobin molecules in the blood. This means that free oxygen is always more concentrated in the air capillaries than in the blood, so more oxygen goes into the blood from the lungs than vice versa.

The concentration of carbon dioxide in the blood is always much higher than it is in

(⌄) Without abundant plant life releasing oxygen into the atmosphere, birds and other animals could not exist on Earth.

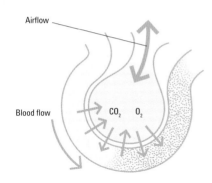

Airflow

Blood flow

CO_2 O_2

⊕ Gas exchange between an air capillary and a blood capillary works through diffusion.

the air capillaries, as it is constantly being generated by body tissues, so carbon dioxide passes from the bloodstream into the lungs much more readily than in the other direction. Excess carbon dioxide in the blood (hypercapnia) can produce a range of harmful effects in the body, including blood acidosis; being in this condition quickly triggers a bird to breathe more rapidly.

The most abundant gas present in atmospheric air – nitrogen – does not react with any compound in blood or other tissues, so while it can diffuse freely across cell membranes, its concentration in the blood is not significantly different from that in the air inside the lungs.

EMERGENCY RATIONS

When a bird's muscles are working exceptionally hard, their demand for oxygen to carry out the energy-releasing reaction may exceed the oxygen supply. In this situation, energy can be released from glucose without oxygen (anaerobic respiration) but this reaction releases lactate (an organic molecule produced in high concentrations in muscles) as a by-product, plus a free hydrogen ion. The buildup of these hydrogen ions makes the muscles' pH more acidic, causing a burning sensation. Therefore, exclusive anaerobic respiration cannot be sustained for very long before aerobic respiration has to resume.

ADAPTATIONS AND SPECIALISATIONS

Changes in atmosphere, as well as a bird's physical tasks, can present a challenge in keeping the body supplied with sufficient oxygen. Birds cope with these issues (up to a point) through behavioural changes, and some species show permanent adaptations to deal with particular conditions.

Like the heart rate, the breathing rate of birds will change according to their needs. The body's demand for oxygen rises dramatically when the bird goes from being inactive to taking flight, and breathing rate increases accordingly, as does heart rate. Breathing rate is also important for heat regulation – birds do not have sweat glands, so panting is their main way of losing excess heat. When a bird pants, it loses heat through water evaporation from the lungs via the open mouth, so it must drink more water to avoid dehydration.

Some birds have permanent adaptations to help them survive in low-oxygen environments. Such species include resident upland specialists like Ptarmigans, which live above the tree line in montane regions, and migrants like Bar-headed Geese and soaring birds like Rüppell's Griffon Vultures (*Gyps rueppelli*), which habitually fly at very high altitudes. These birds show various anatomical and physiological traits that help

⊼ Deep-diving birds can carry more oxygen in their blood than land-bound species.

them extract maximal oxygen from the air they breathe. They have larger lungs, their haemoglobin is quicker to take up oxygen, the membranes between their air and blood capillaries are thinner, allowing quicker gas diffusion, and their muscles (skeletal and cardiac) are more efficient at oxygen take-up.

⟨ The male Trumpet Manucode's ordinary-looking body hides an extraordinary coiled and elongated trachea used to generate its loud calls.

These adaptations are, of course, built upon a breathing system that is already extremely efficient – even non-specialised birds appear to be much better at coping in low-oxygen atmospheres than similar-sized mammals.

Birds that dive have to contend with extended periods of not breathing at all, and their adaptations include a larger blood volume with a richer supply of haemoglobin, to carry extra oxygen. For example, the Tufted Duck (a diving duck) stores about 70 per cent more oxygen in its body than the Mallard (a dabbling duck), and diving birds can also tolerate a higher concentration of carbon dioxide in the blood than non-divers. When underwater, the heart rate slows and some other body parts, such as the brain, reduce their demand for oxygen to help conserve the supply.

(A) Rüppell's Griffon Vulture has been recorded flying at a height of 37,000 feet (11,300m).

TRUMPETING CALLS

Cranes, famous for their loud bugling calls, have elongated tracheae that help them to produce more resonant sound. This adaptation is present in some other birds too, and reaches its most extreme form in the Trumpet Manucode, a bird-of-paradise native to New Guinea and Australia. The male's trachea is about three times as long as the total body length, and is arranged in concentric coils within the chest. This bizarre structure enables the bird to produce calls that carry many miles across the rainforest, to attract the females (which have normal-sized tracheae).

(>) Calling in a 'duet' is an integral part of the courtship ritual in cranes, which form long-lasting pair-bonds.

8

THE DIGESTIVE SYSTEM

Being highly mobile makes birds adept at finding food, but for those that fly, the need to remain light enough to take to the air has implications for the way birds eat and digest their food.

> Waxwings can consume berries at a prodigious rate in winter – up to three times faster in freezing temperatures than in milder conditions.

THE BIRD'S DIGESTIVE SYSTEM

Birds eat organic matter of various kinds, and build their own body's cells and derive their energy by breaking down and then reassembling that food. The digestive process involves several steps and several connected organs, as the organic material is broken down in different ways.

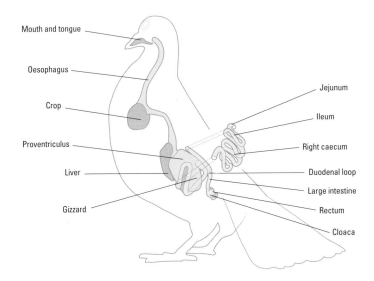

Mouth and tongue

Oesophagus

Crop

Proventriculus

Liver

Gizzard

Jejunum

Ileum

Right caecum

Duodenal loop

Large intestine

Rectum

Cloaca

The gastrointestinal tract is effectively one long tube, though with distinct parts along its length that have particular functions. Its overall function is to break down whatever the bird eats, extract everything of value, and eliminate the rest.

Because birds lack teeth, and most do not have a particularly large or mobile tongue, food items are often swallowed intact or in large pieces, passing down the oesophagus or food pipe to the crop. The oesophagus, like much of the rest of the digestive tract, is lined with smooth muscle, which pushes food downwards through waves of contraction (peristalsis). The crop, an outpouching of the oesophagus rather than a stomach as such, moistens and softens swallowed food. It is

⊗ The digestive tract of a pigeon.

often very large, and highly expandable. Birds that regurgitate food for their young, such as pigeons, store food here; it is also a vital food store for birds that need to consume a lot of food quickly, such as vultures, with their unpredictable feeding opportunities, and hummingbirds, which are often chased away from nectar-rich flowers by fiercer rivals. However, some birds have no crop – among them are geese, owls, and buttonquails.

Food breakdown begins in earnest in the proventriculus, a small glandular chamber in the tract, leading on from the crop, where protein-digesting enzymes are released.

From there it proceeds to the gizzard, a muscular organ that breaks the food down through its powerful contractions. These two parts of the tract are analogous to the human stomach.

THE INTESTINES

Beyond the gizzard are the loops of the small intestine, a tubular structure with tiny finger-like projections (villi) lining its inner walls. The small intestine has three regions: the start of it is known as the duodenum, the middle stretch is the jejunum, and the last part is the ileum. In the duodenum, quite close to where it exits the gizzard, the food is mixed with secretions from the gallbladder, which collects and stores bile produced by the liver, and also from the pancreas, which releases food-processing enzymes.

By the end of its transit through the small intestine, the food has been reduced to a fluid with little remaining in the way of useful nutrients, and the very short large intestine or colon mainly just absorbs excess water.

(ᐱ) Vultures can store large quantities of meat in their crops, which means they can eat quickly at a carcass before larger scavengers drive them away.

In most birds, two narrow pouches extend from the colon; these are the caeca, which have a role in breaking down some food types and absorbing certain salts as well as water. Waste exits the body via the bird's single excretory (and also reproductive) opening, the cloaca.

FOOD PROCESSING THROUGH THE TRACT

Within a bird's body, the processes involved in building new cells and tissues and maintaining the various internal cycles all take place at the microscopic, cellular level.

Only very small, simple molecules can enter and leave cells, so food items need to be broken down to this level through the digestive process before they can be used. This involves both mechanical and chemical processes. In most birds, the mouth secretes saliva, which moistens the food and makes it easier to swallow. In some birds, it contains amylase, the first enzyme the food will meet. Enzymes are protein molecules that break down the chemical bonds holding molecules together, by 'locking on' to specific sites on the molecule. Salivary amylase breaks large carbohydrate molecules into smaller pieces; this process continues after swallowing. Many birds also produce mucus in the mouth to help swallowing, and swiftlets use this mucus as a nest-building material.

Some food items may be torn up or squashed somewhat before they are swallowed, but in general, much less mechanical processing occurs in the mouth than is the case with mammals. The crop does not release enzymes, but does secrete a fluid mucus, which softens the food by dissolving soluble parts of it.

Between the proventriculus and the gizzard, the bulk of early food processing work takes place. In the proventriculus, acids and proteases (protein-digesting enzymes) are secreted into the food, and the gizzard's vigorous contractions break it down mechanically. The food is periodically squeezed back into the proventriculus for a further dousing with enzymes, then returns to the gizzard. Birds that eat very hard foods,

(ⱽ) The white content of bird droppings is primarily uric acid filtered from the blood by the kidneys.

such as grains, often swallow grit and other stones, which stay in the gizzard and help to grind up the food further. Large indigestible items like bones and fur are compacted into pellets in the gizzard and regurgitated (see pages 138–39).

At the start of the duodenum, bile from the liver is secreted via the gallbladder. Bile salts have an emulsifying action – they collect around droplets of lipids (fats) in the food and prevent them from joining together into larger particles. These droplets are broken down by lipases – fat-digesting enzymes released into the small intestine from the pancreas. Pancreatic fluid also contains proteases and amylases, the enzymes that act on proteins and carbohydrates respectively.

THE FINAL STAGES

Through the actions of the enzymes, carbohydrates are broken down into glucose and other simple sugars, proteins into amino

(A) In winter, small birds' survival chances are improved if they have access to protein-rich foods such as nuts.

acids, and fats into fatty acids. These molecules are small enough to be absorbed into the blood, via capillaries in the intestinal villi. They are then transported in the blood to where they are needed, or can be stored (for example, excess glucose is stored in muscle tissues and in the liver, and when these stores are full it is converted to fat and stored in adipose tissue).

Besides protein, fats, carbohydrates, and water, food contains small amounts of other important micronutrients, such as vitamins and minerals. These are absorbed in the small and large intestines. Food content that the body cannot use (including certain nutrients, if present in excess) is excreted. Bird faeces also contain uric acid filtered out by the kidneys, and excess water, as birds do not have a bladder or separate urinary opening.

BILL AND TONGUE ANATOMY

Lacking forelimbs capable of manipulating objects, birds carry out almost all food-handling with their bills and tongues, and the diversity of their diets is reflected in the bill sizes and shapes found across the avian world.

A bird's bill is a sensitive and sometimes quite flexible structure, formed by the rostrum (upper jaw and nasal area combined) and the mandible (lower jaw). It is made of bone, sheathed with a layer of tough keratin, which continues to grow and to wear down.

Among passerines, two bill shapes predominate. Insectivores like warblers usually have quite long, thin, and delicate bills, often with a slight down curve. They are made for speed and delicacy rather than power, to feel into small spaces and snap on to food items that might move quickly but are small and soft enough to be easily managed. Seedeaters like finches have conical, thick-based bills with sharp cutting edges and a strong bite, suitable for crushing seeds so that indigestible husks can be discarded. Omnivores, such as vireos and thrushes, have an intermediate bill shape – slim enough to probe but strong enough to squash fruits and softer seeds.

Among other bird groups, bill shapes include the hook tips of birds of prey, to tear gulp-sized pieces of meat from large prey that is held down with the feet, and the sensitive flattened bills of ducks and some other aquatic birds, to filter food items from water. Sea ducks like scoters have thick-based bills and use crushing actions with the bill and manipulation with their relatively large and thick tongue to discard the shells of their mollusk prey before swallowing.

TONGUE VARIETY

Most birds have a relatively small, simple, pointed tongue that fits into the lower mandible. It is supported by small bones, and has some backwards-pointing spines near the base to aid swallowing. Exceptions include the parrots, which use their thick, round-tipped tongues almost like fingers, to move food items around in their mouths; the tongue is also involved in sound production. Lorikeets, which are nectar-eating parrots,

◀ Penguins have strong, spiny tongues to manage their slippery prey.

Ⓐ Saw-billed ducks like the Red-breasted Merganser have tooth-like serrations in their bills to grip fish.

Ⓐ Birds that kill and eat vertebrates have hooked bills, to tear up their large, tough-bodied prey.

Ⓐ Thrushes are omnivores with general-purpose bills that can handle invertebrates and also fruits and softer seeds.

Ⓐ Seedeaters' bills are conical – fine at the tip to pick up seed, and thick at the base to crush it.

have brushlike tongue tips to soak up their liquid diet. Hummingbirds' long tongues are fork-tipped, the tips coming together to trap nectar. The upper part of the tongue then bends, which generates a pumping action that pulls nectar into the mouth. Woodpeckers use their extremely long, spiny, and sticky tongues to collect insect prey from deep crevices in tree bark and from ant burrows in the ground. Penguins' tongues are entirely spiny, providing extra grip on their slippery prey as they swallow.

Ⓐ A Green-winged Teal uses its flat, lamellae-lined bill to filter small items of food from water.

OTHER DIETARY ADAPTATIONS

There is very little plant or animal material not eaten by at least some species of birds. Even notoriously hard-to-digest foodstuffs like tough grass blades or decaying meat are the staple diet of certain birds. Their digestive systems show a range of adaptations to cope.

Organic matter is made up primarily of carbon, hydrogen, and oxygen. Atoms of these elements are held together with chemical bonds, in molecules of various types. Some molecules are chain-like, some ring-shaped, and many are large and complex, incorporating both chains and rings, as well as additional twists and folds. The strength and number of the bonds affects how easily they can be broken down into small, simple molecules.

Grasses contain the tough carbohydrates cellulose and lignin. Most animals cannot digest these compounds at all, but birds that feed on grass, such as geese, are thought to be able to break it down to some extent thanks to fermentation by specialised bacteria present in their exceptionally long

Ⓐ Grebes feed their chicks on fish and also feed them feathers to help protect their insides from sharp spines and bones.

Ⓥ The King Vulture's digestive tract is powerfully acidic to remove harmful bacteria from past-its-best meat.

caeca. Bearded Vultures are among the few birds that can digest bone, thanks to very powerful stomach acid, and indeed the species' diet may be up to 90 per cent old, dried bone pieces.

All birds have a community of gut bacteria, which provides various benefits, including helping with the breakdown of food and removing toxins. Some of the toxins that may be present in the digestive tract are produced from other, non-resident bacteria that the bird has consumed. Carrion-feeders like the New World vultures eat a large quantity of material that has already begun to decay and contains potentially harmful bacteria. To compensate for this, the vultures' digestive system is acidic enough to kill off most of these bacteria, and these birds also have a much higher tolerance than most to the toxins released by pathogenic bacteria.

Hummingbirds, feeding primarily on nectar, have an extremely rapid digestive transit – the crop empties fully in just 15–20 minutes (this takes more than an hour in most birds). Nectar moves directly from the crop to the small intestine – only solid objects like tiny insects spend time in the stomach for processing – and becomes available as a fuel source almost immediately. The entire digestive process takes less than an hour, and some 97 per cent of the sugars consumed are absorbed.

FEEDING CHICKS

Some of the foods swallowed by birds could harm them; this is a particular risk for very young birds. When collecting food for their chicks, insectivorous birds will decapitate large beetle larvae that might attempt to chew their way out of the body if swallowed alive. You may see an adult grebe feeding soft feathers to its chicks along with whole small fish – the feathers are thought to help protect the young bird's digestive tract from being damaged by sharp fish bones or spines.

UNUSUAL DIETS

The bird species that thrive most successfully in all environments are the omnivores. Versatility in diet improves the chances of survival, but there is also room on our planet for some extreme dietary specialists, with specialised anatomy to match.

Specialism in diet is most clearly reflected in a bird's bill shape. Among the finches, one genus stands out – *Loxia*, the crossbills. In these birds, the mandible tips are elongated and crossed over, forming a tool perfect for extracting pine seeds from between the tight-packed scales of pine cones. Crossbills are also adept at using their strong, long-clawed feet to manipulate the cones. Another expert extractor is the Snail Kite, which has a very long, strongly down-curved bill tip for picking snails out of their shells. The Wrybill, a small wader native to New Zealand, has a sideways bend to the tip of its slim, longish bill, which is thought to be an adaptation to probing among gravel for invertebrate prey, in the braided riverbeds where it forages.

Few birds eat leaves, as the cell membranes contain difficult-to-digest cellulose and lignin. The Capercaillie, a large forest grouse, eats only pine needles in winter, and relies on caecal bacteria to help it ferment this tough diet. However, it eats other foods in the warmer months, and the community of bacteria living in its caeca changes seasonally in response to this dietary shift. The Hoatzin, which lives in South American forests, feeds on tree foliage and, like the Capercaillie and

ⓥ The Snail Kite's peculiar bill is an evolutionarily crafted snail-winkling tool.

(◄) Most fish-eaters catch prey with their bills, but the Osprey does it with spiny-soled, long-taloned feet.

(↳) Capercaillies eat pine needles – one of the toughest foods to digest, which they can only break down with the help of fermenting bacteria.

(↳) The Hoatzin smells like cow manure thanks to a similar community of bacteria in its crop to that in a cow's stomach.

scales, and long, strongly curved claws covered with backwards-pointing scales.

A nectar diet delivers pure glucose and other simple sugars, which the body's cells can use for energy very quickly as no further breakdown needs to occur. Nectar-feeders include hummingbirds, sunbirds, honeyeaters, and sugarbirds, while many other small birds occasionally take some nectar. Adaptations to this diet include the hummingbird's 'shortcut' stomach, which allows nectar to pass directly from the crop to the small intestine, with the protein-rich element of the diet being held back and digested in the stomach.

Wide-gaped birds

other leaf-eaters, needs help from an internal bacterial community to break down the cellulose through fermentation. However, in this species the fermentation takes place in the crop rather than the caeca, and gives the bird a very distinctive smell, reminiscent of cow manure, which may discourage predators.

The Osprey is a raptor that feeds only on fish, and unlike most fish-eating birds it catches its prey with its feet rather than its bill. The foot anatomy is distinctively adapted for gripping slippery prey, with short, stout toes covered on their underside with spiky

Swifts and swallows feed on small insects that they catch in flight. They have exceptionally wide gapes to swiftly engulf and swallow what they catch, so they can continue to hunt without interruption. One bird of prey, the Bat Hawk of Africa and South-east Asia, also has an unusually wide gape but takes larger prey – it catches and swallows small bats on the wing. With an exceptionally rapid digestive transit, it can consume several bats in rapid succession, a trait that allows it to fully exploit the brief but bountiful daily hunting opportunity as bats are leaving their roost in the evening.

PELLETS

For birds that feed on other animals, most meals contain some parts that they cannot digest. Rather than painstakingly picking out only the soft, digestible parts of their meal, these birds often swallow the lot and later regurgitate compressed pellets of all the indigestible bits.

Even the most powerful of predatory birds cannot afford to take too long over its meals. There are always other predators and scavengers around that may try to steal the kill. Therefore, birds eat fast and deal with swallowed bone fragments and the like later on. Most raptors (hawks, falcons, and other day-flying birds of prey) tear up their prey to some extent as they feed, and will not swallow larger bones. Owls tend to swallow most prey intact or in larger pieces. The exception is food offered to very young chicks – the parent will tear off tiny pieces of meat. However, within a week or two the chicks are fed larger and

Ⓥ Bones and hair only travel as far as an owl's gizzard before being squeezed into pellets and disgorged.

intact prey items and begin to produce pellets themselves.

When the swallowed food reaches the gizzard, only that which gets broken down to small enough particles will proceed. What remains is pressed together by the gizzard's contractions, squeezed mostly dry, and then regurgitated in the form of a pellet. Typically, an owl or raptor will produce one pellet for each meal, and the pellet will emerge several hours after eating. The bird is not able to feed again until this occurs.

Large bones, like skulls, are often intact within owl pellets – the bone-eating Bearded Vulture will actively seek out owl pellets to supplement its diet. There tend to be only bone fragments in the pellets from other birds of prey, but their pellets do usually include some recognisable body parts, such as feathers and fur. Analysing pellet contents is therefore a good method of determining a bird's diet; the same can be achieved from analysing faeces, but is considerably more difficult. It is often also easy to tell which bird species ejected a pellet, as pellets tend to be quite consistent in size and general appearance.

WHERE TO FIND PELLETS

Pellets of owls and raptors accumulate at favourite roost sites as well as around the nest. They are often held together quite strongly by the fur of the mammals they have eaten and can be found intact weeks after being ejected. Other birds tend to eject their pellets more haphazardly. Pellets from

Ⓐ Because owls swallow their prey whole, their pellets often contain unbroken skulls or other large bones.

Ⓐ Bird pellets become compressed and smoothly rounded in the gizzard so as not to cause damage on their way out.

kingfishers are made of fish bones and scales, those from insect-eaters are made of tough insect exoskeleton parts, and shorebirds produce pellets containing mollusk and crustacean shell fragments. All of these pellet types are difficult to find, as they are dry, small, and break down very quickly.

Some pellets are composed of plant matter rather than animal remains. Rooks are omnivorous and visit wheat-fields after harvest to eat spilt grain. They may then produce crumbly pale pellets made entirely of wheat chaff.

Ⓥ Pellets on the forest floor can give away roosting spots used by birds of prey.

KIDNEYS AND OTHER WATERWORKS

The correct balance of water in the body is crucial for a bird's survival, and because they fly and must, therefore, maintain a low weight, carrying large volumes of liquid is not possible. The avian urinary system helps to keep the fluid balance correct.

Water is constantly lost through evaporation in the breath, and through excretion, so must be replenished by drinking. Excess water, though, can over-dilute the blood. The bird's body is therefore constantly removing and reclaiming water, as required to maintain balance, from the blood and digestive tract.

Like other vertebrates, birds filter their blood through their two kidneys, which retain the blood's cells, the useful molecules carried in the bloodstream, and most of the water content, while extracting uric acid (a by-product from the breakdown of protein) and other waste products for excretion. Avian kidneys are long, narrow organs with a lobular structure, and they sit above the digestive tract. Within the kidney's cortex (outer layer) are numerous nephrons. Each nephron comprises a hollow capsule connected to a urine-collecting tubule, and a bundle of blood capillaries (glomerulus) within the capsule. Blood plasma passes through the membranes of the glomerular capillaries into the capsule, and then electrolytes and other metabolites are filtered back in, along with some of the water. What remains in the tubule passes through to a collecting duct.

This urine passes along the collecting ducts to eventually reach the ureters, two long pipes that connect the kidneys to the

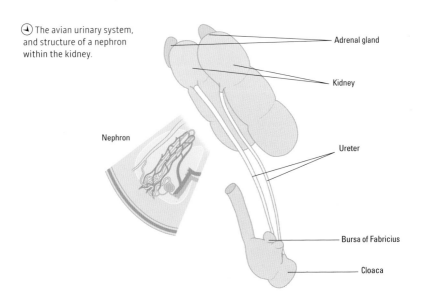

④ The avian urinary system, and structure of a nephron within the kidney.

Adrenal gland

Kidney

Nephron

Ureter

Bursa of Fabricius

Cloaca

⊙ Water is heavy to carry around in the body, so most birds drink only small amounts at a time.

cloaca. Urine is expelled into the cloaca and from there moves up into the colon and the caeca. Water within the urine is reabsorbed here, according to the body's needs, while the uric acid is added to faecal matter and excreted. Uric acid makes up the white part of bird faeces.

The reabsorption of water in the colon helps compensate for the fact that birds' kidneys are relatively inefficient. However, birds still generally fare less well than mammals when short of water, although their greater mobility means they are much better able to find a water source than many mammals are.

Birds' kidneys have an additional role, which is forming glucose (to meet energy needs) out of non-carbohydrate molecules, such as amino acids and fatty acids. This process, gluconeogenesis, is of particular importance in birds whose diet is naturally low in carbohydrates.

HOW BIRDS DRINK

Most birds drink by taking water into their mouths, sometimes using the tongue as a scoop, and tipping their heads up to swallow. Pigeons and doves can suck up water directly; water is drawn into the almost-closed bill through capillary action (the tendency of fluids to rise within a narrow tube) and pumped into the throat with pushing motions of the tongue. Seabirds that drink seawater can excrete excess salt via the salt glands in front of their eyes (see page 107).

ⓐ Albatrosses' bills frequently drip salty water excreted from the salt glands.

THE LIVER AND SPLEEN

Some of the organs in the body are not readily assigned to a single system but are associated with several, and fulfil multiple functions. This category includes two major abdominal organs – the liver and the spleen.

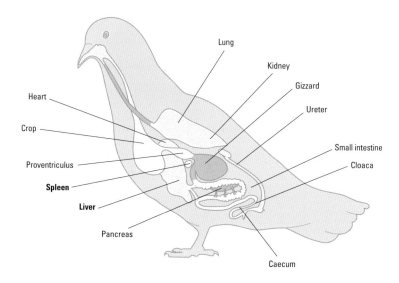

Lung
Kidney
Gizzard
Ureter
Heart
Crop
Small intestine
Proventriculus
Cloaca
Spleen
Liver
Pancreas
Caecum

The liver is a large and prominent organ that sits in the abdominal cavity behind the heart. It is a darker red-brown colour than other internal organs, except in newly hatched birds, in which it is yellow. The avian liver has a less distinctly lobed structure than the mammalian liver, though it does have two distinct halves, each with its own bile duct leading from the gallbladder to the duodenum. It receives blood from the digestive tract via the hepatic portal vein, and one of its major functions is to filter out and store glucose from the bird's last meal. It also breaks down potentially harmful substances from the bloodstream into less dangerous compounds.

The long list of bodily tasks carried out by the liver also includes clearing bilirubin (a

Ⓐ The principal internal organs of a typical bird. Although not directly part of the digestive tract, the liver and pancreas have vital functions in the digestive process.

by-product from the breakdown of old erythrocytes, which is harmful when it accumulates to excess) from the blood; producing bile for the metabolism of fats; storing fat-soluble vitamins, iron and copper for later use; producing blood plasma proteins, and synthesising fats (building them from fatty acids) and certain hormones. The avian liver is proportionately larger than the mammalian liver. Most of its constructive functions are carried out by hepatocytes, which are relatively large, complex cells containing many organelles, while Kupffer cells, which are phagocytes (cells that

deactivate harmful cells and molecules by engulfing them), are the main sites of cell and toxin breakdown.

 Gulls foraging on rubbish tips are often affected by botulinum poisoning.

The spleen

The spleen in birds is a fairly large, pulpy organ, rounded or pyramidal in shape depending on species, that lies beside and to the right of the proventriculus. Despite its proximity to the gastrointestinal tract, it is not involved in the digestive process, but in blood filtration and the immune response. It has been shown to change size quite considerably between the seasons in at least some bird species, for reasons that are still unclear.

In mammals, the spleen has a number of documented functions. It is the maturation and storage site for antibody-carrying lymphocytes for the immune system. It is an important storage area for erythrocytes, and also filters out old and damaged erythrocytes and platelets, and stores the useful by-products from this process before returning them to the bone marrow. It manufactures opsonins, compounds that aid the immune response by 'marking' antigens on cell surfaces so that white blood cells will target them. However, the function of the spleen in birds is not as well-studied as in mammals and function may be different – for example, the spleen does not appear to store erythrocytes in birds.

9

THE REPRODUCTIVE SYSTEM

Some of the most appealing traits of birds – their bright colours, song and the nests they build – are consequences of the biological imperative to find a mate and reproduce.

⊙ The anatomical differences between male and female birds drives their different reproductive behaviours. Mating can literally be a delicate balancing act.

MALE AND FEMALE REPRODUCTIVE ANATOMY

Birds, like most other sexually reproducing organisms, have two sexes. In some species the sexes look alike; in others, extremely different. However, there is little interspecies variation in internal reproductive anatomy.

Males and females produce different types of gametes (the 'sex cells', which when united form an embryo). The male sex produces sperm – small, freely moving and abundant gametes – and the female produces ova or eggs, which are large, cannot move on their own, and are formed mainly of yolk, with a white spot (the germinal disk) on the outside, which holds the cell nucleus. Ova are produced in much smaller numbers than sperm. Each gamete contains half of the parents' chromosomes, so each embryo inherits half of its genes from each parent.

Sperm and ova are formed in the gonads – the testes in males and ovaries in females. In birds, these are located deep inside the body, close to the kidneys. Male birds have two functional testes, and each has a duct (vas deferens) leading to the cloaca. These open out just below the exit of the ureter (the duct connecting the kidney to the cloaca). Spermatocytes – precursors to sperm cells – are formed within the seminiferous tubules within the testes, and mature into sperm cells. In most birds, sperm passes from male to female when they touch their cloacas together, but in wildfowl, ratites and a few other families, the male's cloaca holds an inside-out organ, equivalent to the mammalian penis, which extends out of the cloaca and penetrates the female. It is short and straight in some species, but in others long and corkscrew-shaped.

In females of most bird species, there is only one functional ovary and oviduct, on the left side – the right-side organs never

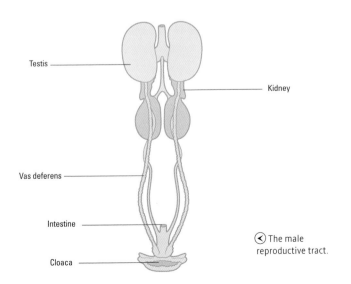

Testis

Kidney

Vas deferens

Intestine

◀ The male reproductive tract.

Cloaca

146

> Bee-eaters, like most birds, mate by bringing their rear ends together in a 'cloacal kiss'.

fully form. Raptors are exceptions, typically having two fully formed and functioning ovaries and oviducts. Although the oviduct is a single continuous tract, the widened part of it where shell is added to an egg is sometimes called the uterus, and the lowest part, just before entering the cloaca, is sometimes called the vagina. The ovary contains between 500 and 4,000 oocytes (the cells that will mature into ova), within follicles. The follicles containing mature ova are much larger than the rest, giving the active ovary the appearance of a bunch of grapes.

SEX DETERMINATION

Whether an embryo develops into a male or female bird is determined by which sex chromosome it inherits from its mother.

In males, the two sex chromosomes are the same (ZZ) while in females they are different (ZW). Each parent passes on one sex chromosome to each of its chicks. The male always contributes a Z chromosome, the female either a Z or a W. Chicks that inherit her Z chromosome will be male (ZZ) and those that inherit her W chromosome will be female (ZW).

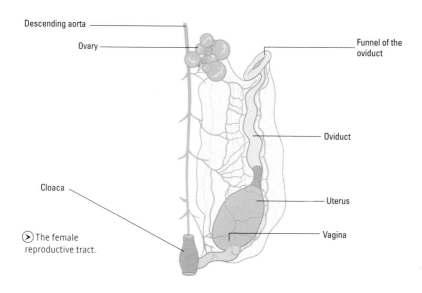

Descending aorta

Ovary

Funnel of the oviduct

Oviduct

Cloaca

Uterus

Vagina

> The female reproductive tract.

MATING SYSTEMS

Swans stay with their partners for life, but permanent monogamy is just one of an array of different mating systems used by birds. How birds conduct their love lives is tied into the type and level of care their offspring require.

Monogamy is much more common in birds than in mammals. Baby mammals develop inside their mother's body and, when born, feed only on her milk. So, in most cases, the father's parental role is limited or non-existent. Birds, though, lay eggs, which need near-constant warmth and, in many cases, the chicks need to be brought food regularly through the day for several weeks after hatching. These tasks are a burden for a lone parent, so having both mother and father involved makes success more likely. Birds that are monogamous often show little or no sexual dimorphism (see page 202), and their courtship often involves a ritualised show of parenting behaviour – the male feeding the female – as well as showing off their physical condition.

Apparent monogamy doesn't necessarily mean fidelity, though. Males and females alike will rarely pass up the chance of an extra-pair copulation, as this means increased genetic diversity in the brood for the female, and the chance of young in more than one nest for the male. Extra-pair paternity occurs in about 90 per cent of the apparently monogamous bird species that have been studied in this respect, and rates can be very high – about 90 per cent of Reed Bunting (*Emberiza schoeniclus*) nests contain chicks that were not fathered by the male bird that cares for them. However, males that observe cuckoldry may respond by investing less effort in their nests, so the behaviour is usually furtive.

⌄ Swans form a life-long pair-bond, becoming more skilled parents with each brood they raise together.

⊙ Female Reed Buntings are not faithful to their nest-partners but conduct their liaisons discreetly.

⊙ Nearly all clutches of Reed Bunting eggs are fathered by at least two different males.

Polygamy

Polygamy, where one bird has two or more long-term partners, occurs in several species. It takes three forms – polyandry (one female pairing with multiple males), polygyny (one male pairing with multiple females), and polygynandry (when both members of a pair have additional partners). All three may be noted within the same species, for example, the Dunnock (*Prunella modularis*), with the best tactic varying according to how easy it is for the birds to find food for their chicks. Polyandry, with three (or more) birds provisioning one nest, is an effective strategy when resources are scarce. Polygyny is more frequent in times of plenty, when a male can afford the time and energy to keep two sets of young birds supplied with food.

Variations on pair-bonding

In birds that produce fully precocial chicks (see page 176), the male is often surplus to requirements after mating, and so no pair-bond forms. Females in these species choose their mates on the basis of body condition, and males compete to show themselves to be best, and thus win as many mating opportunities as possible. In such 'promiscuous' species, there is usually marked sexual dimorphism, with males being larger and more colourful than females. However, there are always exceptions and in this case swans and geese break the rules, being strictly monogamous even though their chicks are precocial and self-feeding from day one. The role of male swans and geese is not to provision but to protect, a task they take very seriously – accordingly, they are larger and more powerful than females, though identical in plumage. In the Black Swan (*Cygnus atratus*), up to 25 per cent of pairs are male–male – they recruit a female to provide eggs, but incubate and rear the brood without her, and their superior ability to see off potential predators makes them more successful parents than male–female pairs.

HORMONES AND BREEDING BIOLOGY

The reproductive cycle comprises a sequence of highly time-sensitive physiological events, regulated by hormones carried in the blood. A complex hormonal balancing act keeps things efficient, as well as effective.

The urge to reproduce is powerful, as it needs to be, but behaviours such as establishing and defending a breeding territory, courtship, nest-building, incubation and care of a brood of chicks all consume a great deal of a bird's time and energy, which can impact significantly on long-term survival. Most birds have a set breeding season; in the temperate areas, the majority of species do not begin to nest until spring, although courtship and pair formation may begin in early winter. Breeding is usually complete by mid-summer. The gonads – glandular organs that, as well as producing gametes, secrete the sex hormones testosterone and oestrogen – become inactive after breeding. The testes and ovarian follicles shrink in size, often dramatically – in the American Crow (*Corvus brachyrhynchos*), for example, each testis is 19 times heavier early on in the breeding season than after it has ended. Changes in day length as spring approaches stimulate the production of gonadotropin-releasing hormone (GnRH) in the hypothalamus, a brain region that can detect light directly, through the skull. The presence of GnRH triggers the pituitary gland to release luteinising hormone and follicle-

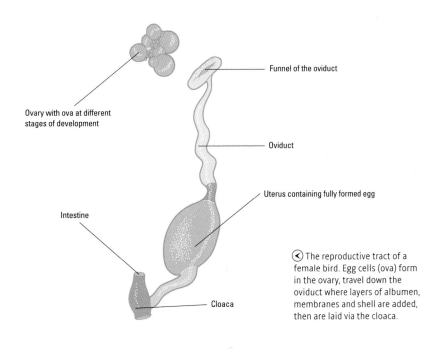

Ovary with ova at different stages of development

Funnel of the oviduct

Oviduct

Uterus containing fully formed egg

Intestine

Cloaca

◀ The reproductive tract of a female bird. Egg cells (ova) form in the ovary, travel down the oviduct where layers of albumen, membranes and shell are added, then are laid via the cloaca.

stimulating hormone (LH and FSH). These two both stimulate the gonads to grow and to start releasing the sex hormones.

As testosterone increases in a male bird's blood, the pituitary gland responds by releasing more LH and FSH, until an optimal level of blood testosterone is reached. Testosterone stimulates sperm production in the testes and also triggers appropriate sexual behaviour (including aggression towards other males). Excess testosterone, though, can be harmful, reducing the effectiveness of the immune system, distracting the bird from normal parental behaviour, and even causing cancer, so fine control of its level is vital. In females, oestrogen released from the ovary triggers maturation of the oocytes into ova and also stimulates female-typical breeding behaviours.

Ⓐ Elderly female Mallards may begin to develop male-like plumage. The dark bill markings reveal their true sex.

AVIAN INTERSEX

A very rare complication can occur in female birds that suffer damage to their ovary and stop releasing oestrogen. This can cause the rudimentary right ovary to begin to develop instead, but, without the feminising effects of oestrogen, it develops as a testis instead, and releases testosterone. The bird therefore begins to develop male-type plumage and show masculinised behaviours – it may even produce functional sperm. This condition is sometimes known as intersex.

This is most often observed in female wildfowl that have lived to old age, either in captivity or a safe wild environment such as a town park.

SELECTIVE BREEDING

People have been breeding birds in captivity for thousands of years. Just as natural selection means that the most 'survivable' individuals tend to reproduce and pass on their traits, people can choose which individual domestic birds to breed together, to perpetuate the traits they prefer.

The most familiar domesticated birds are chickens, ducks, geese, pigeons, turkeys, guinea fowl and canaries. All are descended from a wild ancestor or ancestors still extant – for example, all domestic pigeons and their free-living feral relatives (street pigeons) descend from wild Rock Doves (*Columba livia*), which live in rugged and coastal areas in Europe, north Africa and western Asia.

The process of artificial selection has allowed us to develop strains or breeds of birds with unusual anatomical features not found in their wild relatives. When an unusual genetic mutant appears in a wild population – for example, a pure white pigeon, or a pigeon with a tuft of elongated feathers on its head – natural selection usually makes it less likely to survive. However, such a bird born in captivity will probably not only survive but will pass on its genes at a higher rate than its 'normal' siblings, as bird-keepers appreciate unusual-looking birds and breed from them. If the odd trait is down to a recessive gene (meaning it is only expressed in birds that carry two copies of the gene, and not in 'carriers' that have only one copy), then back-crossing the original oddity to its offspring will produce more birds showing that trait. Over time, different traits can be combined in one lineage, so that a breeder could produce pigeons showing both white plumage and a head crest. Because artificial selection is deliberate and survival of all offspring is likely, it works much more quickly than natural selection.

Through artificial selection, we have developed oversized chickens and ducks for their meat, with greatly enlarged pectoral muscles. The Indian Runner duck (*Anas platyrhynchos domesticus*) has a differently positioned pelvis to wild Mallards, producing a comically erect stance and an ability to run fast on land, making it more easily 'herded' and moved around than other ducks. The Waterslager Canary (*Serinus canaria domesticus*) has a congenital malformation of the cochlea, which impairs its ability to hear high-frequency sounds, and this has caused it to develop a unique low-frequency song.

Consequences

We develop unusual domestic breeds like these to meet a need, satisfy curiosity, or simply appeal to our aesthetic sense, but the birds themselves are often ill-equipped for survival and may have lifelong health problems. However, racing pigeons are an exception to this, as they are bred to be particularly fast, strong and with superior mental abilities. It is no surprise that escaped racing pigeons often thrive in the wild.

⊙ Selective breeding of domestic birds has propagated and exaggerated many qualities that we value in them, including pure white domestic pigeons, yellow rather than wild-type green canaries, chickens that lay extra eggs, guinea fowl and turkeys that are more tractable than their wild ancestors and flightless Indian Runner ducks that are quick on their feet.

MATING AND CONCEPTION

Selecting a mating partner, especially for monogamous birds, is difficult and time-consuming, but copulation – the process whereby the male inseminates the female – is often extremely brief and undignified, even after a prolonged, tender courtship.

Through courtship rituals, the potential pair assess each other's fitness, food-provisioning and nest-building abilities, and other important traits. Once the choice has been made, mating is just a formality (though one that may be repeated many times during the days before the first egg is laid). In species where there is strong competition between males and no lasting pair-bond, copulation itself may be a more elaborate process.

To copulate, the male bird must balance on the female's back, sometimes gripping her neck feathers for stability. He must then tilt himself backwards while she twists her tail to

(∨) A female Mallard cannot stop a determined male from mounting her, but she can prevent his sperm from reaching her eggs.

one side and raises her rear end – in this way the two bring their cloacas into contact and then the male ejaculates. This 'cloacal kiss' is rapid and easy to get wrong, so repeated matings often occur in quick succession.

In those species where the male has a penis, the male ejaculates after penetration. The female Ostrich lies on her belly to mate and the (much larger) male stands astride her, so has no balance problems to contend with. Wildfowl usually mate on the water, again solving the balance problem. In some wildfowl species, forced copulations are possible – this behaviour is commonly seen in Mallards, as males assault females that have not yet begun nesting in spring. The female cannot prevent the mating but she does have several blind-ended extra passages

⊙ The Ostrich female lies down when she is ready to mate – no other position could work for these long-legged, flightless birds.

in her vagina, which, if inseminated, will not fertilise her eggs, so she has some control over which males father her young. The Dunnock, another species in which competition between males is intense, demonstrates sperm competition – a male will peck at the female's cloaca before copulating, to encourage her to eject any sperm from recent previous matings.

OVUM TO EMBRYO

After the male has ejaculated, the sperm cells travel to sperm storage tubules at the base of the uterus, though only a small percentage of them are successfully stored – most will be lost the first time the female defecates after copulation. Once an ovum has been released from the ovary and collected into the oviduct, the stored sperm are carried upwards to the top of the oviduct. They swim over the surface of the ovum, using chemical cues to find the germinal disk. Several may enter the germinal disk, but only one will fuse with the ovum's nucleus to transform it into a zygote – a single cell containing a full set of chromosomes in its nucleus. The genes that make up these chromosomes are contributed equally by its mother and its father, so the young bird is destined to exhibit a combination of its parents' traits. It may also acquire a few new genetic mutations of its own, if there are slight errors in the chromosome copying process when the zygote starts to divide.

Within hours the zygote has begun to divide, and by the time the egg is formed and laid, it has become an embryo, made of many cells that are already differentiating into different types of tissues.

A female bird that has an egg developing inside her oviduct is sometimes described as pregnant, though there are few outwards changes to indicate that this is the case. When ready to lay, she will stay close to the nest and may show signs of restlessness or discomfort for a short time. But being 'with egg' does not impede her ability to move around or to fly if necessary.

FORMING AND LAYING EGGS

Every bird species lays eggs rather than becoming pregnant, enabling them to nurture their embryonic chicks without carrying them around. That means breeding females never lose their ability to fly.

The mature ovum that leaves the ovary already contains its yolk. After fertilisation has occurred, the yolk's outer layer develops and gradually forms an extended spiraled string at either end of the yolk, because the egg turns over and over on its way through the oviduct as layers of albumen (egg white) are added. These two strings, the chalazae, help hold the yolk in a steady and central position, so that it can be fully surrounded and supported by the albumen on all sides.

Albumen provides the embryo with enough water to sustain it through the incubation period, as well as some of the proteins the growing embryo needs. It also acts to slow down transmission of temperature change to the interior of the egg and cushions the embryo and yolk when the egg moves. This includes movement on

(Ᵽ) Eggs do not need to be incubated as soon as they are laid – development of the embryo halts until incubation begins.

its transit to the outside world, and also later on while it is in the nest (the adult birds turn the eggs over frequently during incubation).

The albumen's two membranes are added just before the egg reaches the uterus. Within this chamber of the oviduct, the albumen takes on more water through the membranes, until it is pressed tightly against the uterine walls, forming its ovoid shape. Now the eggshell forms, and this is by far the longest part of the egg-forming process, taking place over many hours, if not a full day. Once this process is completed, the sphincter muscle at the bottom of the uterus relaxes and opens, and the egg can pass down into the vagina and from there to the cloaca.

LAYING THE EGGS

From start to finish the egg-producing process takes about one day, though this varies between species. Most birds will lay one egg per day until their clutch is complete (most often two to six eggs, although in some species ten or more are produced), but in some larger species the interval between each egg may be longer. In most cases, birds' eggs are not very large relative to body size, and the process of passing the egg is usually straightforward. Complications can arise when the mother bird has health issues – for example, a calcium deficiency can result in soft-shelled eggs and the risk that they break inside the bird's body. If a female bird loses her nest soon after one of her ova is fertilised, her body may reabsorb the egg if development is at an early stage.

The kiwis, which lay very large eggs (six times bigger than would be typical for a bird of their body size), have a harder time than other birds when it comes to forming and laying their eggs. The process of egg formation takes 30 days rather than one, including the deposition of an extremely large yolk (65 per cent of the total egg volume, rather than the usual 35–40 per cent). The female's other organs are so squashed by the monster egg that she cannot feed in the last days before she lays. However, she usually manages to lay her egg without difficulty.

ⓥ Large birds such as herons may have an interval of two or three days between laying each egg in a clutch.

BROOD PARASITES

Breeding puts immense stress on a bird's body. It is therefore unsurprising that some species hand over the business of incubating eggs and rearing chicks to unwitting foster parents. This is called brood parasitism, a tactic that has evolved independently in several bird families.

Brood parasitism saves lots of energy and can increase the reproductive output of the parasites – but this is at the cost of the hosts' breeding success. Host and parasite are therefore in an evolutionary battle – the former to continue the deception, and the latter to resist it. Brood parasitism in its simplest form is 'egg-dumping'. Female birds with nests of their own sometimes lay an egg or two in the nest of another of their own species or a closely related one; the behaviour is particularly frequent in ducks but is documented in many others, including gulls, oystercatchers, bluebirds, starlings and sparrows. The behaviour may occur if the female loses her nest before she can lay her egg, but may also be intentional – having young in more than one nest increases the chances that some will survive to adulthood.

Obligate brood parasites are those birds that will only reproduce in this way, laying their eggs in the nests of certain preferred host species. They include some cuckoo species, also cowbirds and honeyguides. Typically, the female of the parasite removes one egg from the host nest and lays one of her own in its place, and visits many nests of her preferred host species over the breeding season. The parasite chick may then kill its foster-siblings, or grow up alongside them (but invariably outcompeting them in the battle to secure food from the host parents).

Cunning cuckoos

Brood parasites are usually considerably larger than their host species, and so lay disproportionately small eggs for their size,

① Nesting Reed Warblers (*Acrocephalus scirpaceus*) are always looking out for Cuckoos, but they have to leave their nests occasionally.

② The Cuckoo egg is a little larger than the host's but does not stand out much at a glance.

to match the host's eggs. This enables them to lay their egg very quickly. In most cases, the egg resembles the host's in pattern as well as size. The Cuckoo (*Cuculus canorus*) has dozens of recorded host species, but each individual female specialises in just one. The cuckoo chick will grow up to be about six times as heavy as its host parents; invariably, it disposes of its foster-siblings soon after hatching, as it cannot afford to share the food brought to the nest by its host parents. It instinctively hoists all other chicks or eggs in the nest on to its back, one by one, and then arches its body to tip them out of the nest.

The Great Spotted Cuckoo (*Clamator glandarius*) uses as its host the Magpie (*Pica pica*). The two species are similar-sized, and the cuckoo chick does not kill off its host siblings but grows up alongside them. There

is evidence that, if the magpie finds and ejects a cuckoo egg, the mother cuckoo will return and take the magpie's eggs from the nest, meaning that breeding success for the magpies is higher if they accept the cuckoo than if they do not. This 'mafia hypothesis' would explain why the Magpie, a notoriously intelligent bird, does not usually remove cuckoo eggs from its nests.

Brood parasites face the additional challenge of developing species-typical behaviour without a same-species 'model'. For example, young cuckoos must migrate to Africa with no experience and no guidance, long after all the adults of their species have already left. They can achieve this purely on instinct, though studies show that once they are adults they refine and improve their route through experience.

③ The baby Cuckoo pushes the eggs out of the nest as soon as it hatches.

④ The adult Reed Warblers appear to be unable to ignore their lone, gigantic foster chick.

STRUCTURE OF A BIRD'S EGG

The egg's components exist to nourish and protect the embryo as it grows from a single cell to a fully formed chick, curled up tightly inside the shell and ready to hatch.

An ovum, released from the bird's ovary into the top of the oviduct, is a very big single cell with its nucleus on the outside in a tiny pale spot called the germinal disk. After fertilisation, this nucleus will develop into the embryo. The rest of the ovum is filled with yolk – the embryo's food store.

Yolk is a yellow, orange, or red fluid, rich in fats and proteins that are synthesised in the mother bird's liver and transported to her ovaries via the bloodstream. It also contains antibodies to kick-start the embryo's immune system. Its colour comes from carotenoid pigments, processed from the food the bird eats. Yolk also contains hormones, which may have lasting impacts on the embryo's eventual fortunes – studies show that male gulls hatched from eggs with high testosterone in the yolk are, as adults, more aggressive in territorial defence.

After fertilisation the yolk is enclosed in a vitelline membrane, to which a new layer, the chalaziferous, is added soon after as the egg moves down through the oviduct. The chalaziferous layer is a coat of protein fibres, which form the two twisted supporting

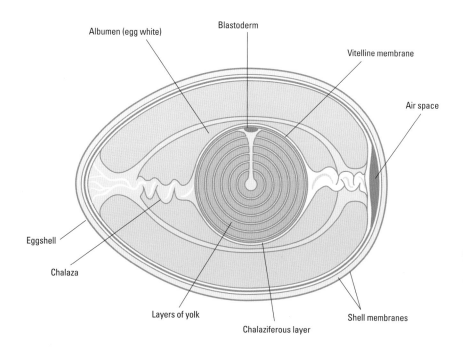

Albumen (egg white)

Blastoderm

Vitelline membrane

Air space

Eggshell

Chalaza

Layers of yolk

Chalaziferous layer

Shell membranes

strings that you may notice extending from the outside of the yolk into the albumen in a cracked chicken egg. In an intact egg, the chalazae stretch towards the ends of the egg, reaching the double membrane between the albumen and eggshell.

The variable amount of protein in albumen alters its consistency. It is deposited in distinct layers. Closest to the yolk it is very fluid, but is surrounded by a thicker, springier layer. The outermost albumen layers are more fluid again. When you look at a cracked egg, the distinctions between the thick and thin albumen types are quite obvious. The albumen's double membrane is separated at the egg's round end, where there is a substantial space or air cell between the inner and outer layers. The hatching chick breathes from the air cell in between breaking the albumen membranes and cracking a hole in the actual eggshell.

Ⓐ Most birds' eggs have a pale background colour, but those laid in open nests typically have dark markings for camouflage.

THE EGGSHELL

The eggshell is made from calcium carbonate crystals, which are deposited on a supporting matrix of proteins and starches, with tiny pores through which air and water molecules can pass. Calcium for egg formation comes mainly from food but up to 40 per cent of it is drawn from the mother's bones. Pure calcium carbonate, a chalky substance, gives the eggshell a white colouration. Pigmentation may also be added at this stage, to give camouflaging colour or pattern to the eggshell in species that make their nests in the open.

Ⓐ The internal structure of a chicken's egg.

VARIATIONS IN EGGS AND EGG CARE

Birds' eggs vary between species in their shape, colour and relative size. These variations suit different nest types and lifestyles, and the various ways that birds care for their eggs.

Some eggs are almost spherical, while others are much narrower and taper to a pointed end. Pointed eggs lose heat faster than rounder ones, having more surface area relative to their volume, and this effect is most pronounced in the smallest eggs. So, in terms of incubation, the larger and rounder the egg, the better. However, carrying around and laying a heavier, bigger and rounder egg is harder work and requires broader pelvic bones and reproductive organs, and birds cannot afford to compromise their mobility too much. Therefore, those species that are powerful fliers and spend much time on the wing tend to lay narrower and more tapered eggs, while the species that produce rounder eggs, such as owls, trogons and pittas, are much less active fliers, with accordingly less streamlined pelvic anatomy.

The default colour for a bird's egg is white, but in nature pure white eggs are relatively rare, except among hole-nesting birds like kingfishers, whose eggs are concealed from predators. Among the rest, an array of colours and patterns has evolved, many of them beautifully cryptic. Eggs laid in well-hidden nests may not require camouflage but still benefit from pigmentation for other reasons. Bright blue eggs, like those of the American Robin, offer protection against ultraviolet light for the embryo. Nests in very open habitats are at risk of overheating so tend to be pale, often with a fine dark pattern for camouflage, such as that on the pale shells of Yellowhammer (*Emberiza citrinella*) eggs. Grebes hide their eggs with wet vegetation when they leave the nest – over time, this stains the shells, providing camouflage.

Almost all birds incubate their eggs by sitting on them, applying warmth from their

Malleefowl use the heat generated by decomposing vegetation to incubate their eggs.

Ⓐ The Long-eared Owl's (*Asio otus*) eggs have a characteristically round shape. Laid in tree holes, they need no camouflaging pigment.

Ⓐ The American Robin's blue eggs absorb less ultraviolet light than unpigmented eggs do.

Ⓐ Guillemots (*Uria aalge*), which nest on cliffs, lay pear-shaped eggs that are unlikely to roll if disturbed by the wind.

Ⓐ Hummingbirds and other streamlined, fast-flying birds lay relatively long, narrow eggs.

skin to raise the internal temperature enough for the embryo to develop. Embryos can survive unwarmed for days after the egg is laid, in a state of suspended animation, but once incubation begins it can only be interrupted for short periods unless the ambient temperature is high.

DIY CHICKS

Malleefowl (*Leipoa ocellata*) lay their eggs on a bed of decaying vegetation, covered with soft earth or sand. This keeps the eggs warm enough for embryonic development. Each male builds and maintains a nesting mound, and one or more females visit to lay eggs once a week or so. The male monitors the nesting mound constantly, adding or removing material to regulate the temperature. Each egg takes 50 to 100 days to hatch. A Malleefowl chick is well-developed on hatching, needing no parental care whatsoever, and once it has dug its way free it wanders away to begin a solitary life.

CAPTIVE BREEDING

A relatively small number of wild birds have been domesticated and kept and bred in captivity for many millennia. These are birds that we eat, keep as pets, or otherwise make use of on a large scale. However, humans regularly breed many more species in captivity for other reasons.

In particular, captive breeding is a valid and much-used strategy in the conservation of endangered species – wild birds are caught and kept safe, and bred in captivity with a view to releasing their offspring or later generations back into the wild. Successful captive breeding requires a full understanding of the species' physiology and behaviour, and also sometimes the willingness to apply creative solutions to specific problems.

The simplest way to breed birds in captivity would be to place a male and female together in a suitable enclosure with a place to nest, provide them with as much food and water as they need, and wait for nature to take its course. This works for quite a few species. Many small songbirds will pair up

and nest even in relatively small cages, and social species can do well if kept in groups in larger aviaries. However, intelligent species that form lasting pair-bonds, such as parrots and corvids, may never accept a mate they did not choose themselves. Even telling the sexes apart may be impossible without DNA testing. Some other species will not settle enough in captivity for normal courtship and nesting to take place. In such cases, artificial insemination can be used, if there is a real conservation need to breed the species in captivity. Tame captive male birds can be

⌄ Captive-reared Cranes (*Grus grus*) at the Great Crane Project in Gloucester, UK. The keepers' grey coats mimic adult Crane plumage as they guide the chicks to find food.

Captive-bred California Condors (*Gymnogyps californianus*) are fitted with trackers when they are returned to the wild so that scientists can monitor their progress.

The Hawaiian Goose or Nene (*Branta sandvicensis*) nearly became extinct in the wild but was saved through captive breeding efforts.

trained to 'copulate' with a semen-collecting vessel, and the resultant semen is then placed in the cloaca of a receptive female.

Foster parenting

Captive birds may choose to ignore the eggs they lay, and if so the use of an artificial incubator is required. This machine maintains the correct temperature and humidity for successful incubation. An alternative to the incubator is to use a foster parent. This method was used to save the Chatham Island Robin (*Petroica traversi*) from extinction. Eggs laid by the sole surviving female of the species were placed in the nests of the closely related Tomtit (*Petroica macrocephala*). The foster parents reared the robin chicks, and the loss of her eggs encouraged the female robin to lay replacement clutches, producing more young at a faster rate than would occur in nature.

Chicks that cannot be raised by their own parents or by avian fosterers must be hand reared. This may be fairly simple or extremely difficult, depending on the species. Obtaining suitable foods is usually easy enough, but many physical needs must be taken into account. Altricial birds in particular are very delicate, needing closely controlled temperature and humidity, very careful feeding to avoid them inhaling liquids, and proper support of the legs and wings to prevent permanent joint damage.

EMBRYONIC DEVELOPMENT

A newly hatched baby bird develops with fantastic speed, but this is nothing compared to the rate and scale of changes that have already taken place while it was inside the egg.

(>) Altricial baby passerines hatch from their eggs in a helpless and featherless state, but their development is very rapid.

EARLY DEVELOPMENT IN THE EGG

An ovum is usually fertilised less than an hour after being released from the ovary. When the egg is fully formed and laid, about a day later, it contains an early-stage embryo.

By this stage, some of the cells are beginning to follow their particular developmental path to become specialised body tissues. Further cell division and differentiation may be suspended after the egg is laid, depending when incubation begins, but as soon as the egg is warmed up to about 20°C (68°F), the embryo resumes development.

The cells in a new embryo are initially in one layer that lies on top of the yolk. As their numbers increase, they form two distinct layers. The ectoderm sits on top and the endoderm below, and soon afterwards the central cells in the layer pull away from the yolk, creating a space into which the embryo will continue to grow. A central layer of cells

(mesoderm) then begins to develop between the ectoderm and endoderm. Cells within each layer begin to differentiate.

The cells in the ectoderm are destined to become the young bird's integument (skin, claws, bill and feathers) and nervous system. Those in the mesoderm eventually form bones, muscles and the blood circulation system and reproductive system, while the respiratory system, glands and digestive tract form from the endoderm. The proteins needed to build new cells come from the yolk and also the albumen, via the yolk.

After a couple of days of incubation, the embryo resembles a tadpole, with a big head and eyes, and tail-like spinal cord. Its heart is already beating, even though it is on the

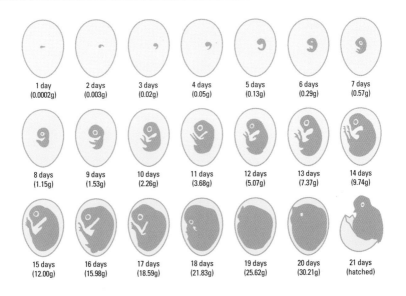

1 day (0.0002g)	2 days (0.003g)	3 days (0.02g)	4 days (0.05g)	5 days (0.13g)	6 days (0.29g)	7 days (0.57g)
8 days (1.15g)	9 days (1.53g)	10 days (2.26g)	11 days (3.68g)	12 days (5.07g)	13 days (7.37g)	14 days (9.74g)
15 days (12.00g)	16 days (15.98g)	17 days (18.59g)	18 days (21.83g)	19 days (25.62g)	20 days (30.21g)	21 days (hatched)

Ⓐ Killdeer (*Charadrius vociferous*) eggs require 24–28 days of continuous incubation before they will hatch.

outside of the body. The embryo forms an external circulatory system (the allantois) to collect nutrients, exchange oxygen and carbon dioxide and dispose of waste. The blood vessels it sends into the yolk are much longer than the embryo itself. If you hold an egg up to a strong light at this stage of incubation, these blood vessels are usually clearly visible as a branching array of red threads from one point of the yolk, even if the actual embryo is barely discernible. Over the next few days the allantois extends to completely surround the embryo.

FIRST SIGNS

The embryo's body starts to curl over and its limb buds begin to appear soon after its heart and eyes are apparent. Other early-stage growth includes the appearance of the mouth and bill, including the egg tooth (a hard button on the bill tip, which the chick will eventually use to break out of the egg), the development of the reproductive tract and the appearance of feather tracts on the skin. The exact timing of these events varies between species, but they all take place in the first half of the incubation period.

Ⓐ The growth of a chicken embryo over its 21-day incubation period.

169

LATER DEVELOPMENT IN THE EGG

Through the second half of incubation, the embryo becomes progressively more bird-like, and its internal systems develop further. As it grows, the chick gradually consumes the yolk and albumen.

Halfway through incubation, the chick's wings have differentiated joints and show the tracts of the flight feathers; the tail feather tracts are also apparent. Its toes and claws are also partly formed. It is capable of movement and its organs (except for its digestive tract) are in their correct positions inside its body. Over the next few days, the first downy feathers will begin to grow on its skin, and the scales start to appear on its legs and toes.

About two-thirds of the way through incubation, the embryo turns its body around to angle its head towards the rounder end of the egg. It will remain in this position until hatching, continuing to grow and to become more and more tightly curled up. Its bill, claws and leg scales become firmer. As the hatching day approaches, the gut, including the yolk sac and what is left of its contents, start to draw into the body cavity, a process that completes about a day before hatching.

The different developmental pace of altricial and precocial chicks becomes most evident in the second half of the incubation period. Purely altricial chicks hatch at an earlier stage of development than precocial chicks – their bones are less rigid, their muscles weaker, their bills and claws much softer and their feather development is negligible. For this reason, the eggs laid by altricial species are usually relatively small compared to the mother bird's body size, as the embryo does not need as much nutrition prior to hatching, and their incubation period tends to be shorter. The Killdeer, a precocial shorebird, incubates its eggs for 24–28 days, for example, while the Common Grackle (*Quiscalus quiscula*), an altricial songbird, has an incubation period of just 11–14 days, despite both species being about the same size when adult.

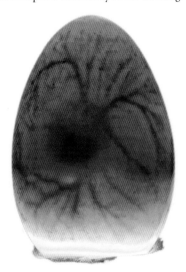

⊙ The blood vessels of the allantois are visible in an egg held up to a strong light. The allantois forms an extensive network outside the body of the embryo, penetrating all parts of the yolk early in the incubation process.

SHARED SPACE

By the time the chick is ready to hatch, there is almost no albumen left in the shell, its fluids and protein having been used up by the developing chick. Throughout incubation, air exchange occurs through tiny pores in the eggshell, supplying the embryo with oxygen, and allowing its waste carbon dioxide to leave the egg. This process also allows the air space or air cell to grow larger as the volume of yolk and albumen shrinks. The air cell is vital to the chick's survival during the hatching process, and by hatching stage it takes up about 25 per cent of the internal space of the egg, with the rest occupied by the curled-up chick.

(A) A hatching chick usually breaks its egg into two neat pieces, with a straight break chipped between them.

(A) Precocial chicks, like this Mallard duckling, are very damp when they hatch but their down quickly dries and becomes fluffy.

DEVELOPMENTAL PROBLEMS

Not every egg will be fertilised, and no amount of incubation can produce a chick from an unfertilised egg, but other problems can also prevent successful development.

I n many mammals, fraternal twins (or triplets, or quadruplets, and so on) are the norm – the female releases two or more eggs at the same time, which are fertilised and gestated together. Identical twins are much rarer and arise when a very early-stage embryo divides into two, producing two genetically identical embryos. This may cause complications – the embryos may not fully separate and conjoined twins are the result. In many cases, though, two healthy young are born, though identical twins are easily missed in a litter of three or more.

In birds, though, the fertilised egg is contained in a rigid eggshell rather than a stretchy uterus, and a normal egg only has space for one healthy chick to develop inside. In domestic chickens, eggs with two yolks are not uncommon. Such eggs are larger than single-yolk eggs. If both yolks are fertilised, two embryos could theoretically grow inside

⊗ Inheriting a second copy of the gene that gives crested Mallards their head adornment will result in fatally deformed chicks.

one egg, and the same is true if a single embryo divides into two, but the chances of both emerging from the egg alive are tiny. Known cases include wild fraternal Eastern Bluebird (*Sialia sialis*) twins observed in a nest in 2013 – both chicks hatched from their oversized egg, but soon died. A conjoined pair of wild identical-twin Barn Swallows (*Hirundo rustica*) was found in 2008.

Some chick deformities are the result of genetic mutation. A dominant gene causes the 'crested' mutation in the domestic Mallard. If a chick inherits one copy of the gene from a parent, its skull does not fully close at the back, and at this point a fatty pad develops from which a pompom-like tuft of feathers grows. However, embryos

(>) Swallows are one of only a handful of bird species in which twinning has ever been documented.

that inherit two copies will have a much more severe skull malformation and will die before hatching.

Domestic chicks and ducklings with additional limbs have been noted on several occasions. These could also be down to genetic mutations, but there is also the possibility of a conjoined parasitic (partially developed) twin.

ENVIRONMENTAL HAZARDS

The pores in the eggshell mean that bacteria can penetrate the egg. A living embryo can deal with this through its immune system, but if the embryo dies early in development, bacteria can cause its decomposition – the result is a rotten or addled egg, famously terrible-smelling if cracked.

Hatching is a huge challenge for the chick. Humidity levels now are crucial – too high

and some fluid may remain inside the egg and cause the chick to drown; too low and the membranes inside the shell may stick to the chick's body and impede its movement.

(v) This significant jaw deformity may not be problematic for a chick developing in the egg, but survival to adulthood is unlikely.

173

INCUBATION AND HATCHING

Incubating a clutch of eggs until they hatch is a hefty commitment, and though incubation looks like total inactivity, it is, in fact, a very stressful time in a bird's life.

The eggs can only be left for brief spells, to ensure they do not become chilled, and are not spotted by a predator. An egg is an easy and highly desirable meal for all kinds of predators and scavengers, and the incubating adult bird is itself quite vulnerable to attack while on its nest. In bird species that are functionally monogamous, both parents are involved in incubation in one way or another. Many birds incubate in shifts – these may last for days in the case of seabirds, with one parent going far out to sea to feed while it is off duty. In many others, the female incubates and the male brings food to her on the nest, only occasionally taking over incubation himself. In such species, only the female develops a brood patch, and she often has superior camouflage to her mate.

'Pipping' – the first crack appears

The chick works to enlarge the crack

The crack is more than half the way around the egg

The chick pushes its back against the crack

Finally it levers itself free

⊙ Hatching is a slow and exhausting process.

⊙ The stages of the hatching process.

If danger threatens an incubating bird, it faces a dilemma – in leaving the nest, it places its eggs at risk but may save itself. Most birds will leave their eggs eventually if a potential predator gets very close, and may then try to see off the predator by mobbing it (the non-incubating parent will join in, as will neighbors within a colony). Sometimes, the incubating parent will try to lure the predator away from the nest by feigning injury – it may run rather than fly away, calling loudly and dragging one wing.

Close to hatching, the parents may leave the eggs for slightly longer. Often the incubating parent will bathe and return to the nest with damp feathers, to ensure humidity is at the right level for the chicks to hatch successfully.

⊙ It takes less than an hour for precocial chicks' down to dry out. Soon they will be ready to leave the nest.

BREAKING FREE

A fully developed chick inside its egg begins the hatching process by breaking the membrane separating it from the egg's air cell. It now breathes air for the first time (prior to this, the allantois handled gas exchange) and its lungs begin to work. However, the air in the air cell is relatively high in carbon dioxide, and this stimulates the chick to jerk its neck, striking its egg tooth against the shell. Once the chick has

⊙ An incubating bird can raise humidity levels in the nest by bathing to moisten the feathers before returning to the nest.

broken through the shell (pipped), it rests and regains energy before continuing to enlarge the initial crack. Then it forces the shell apart by pushing with its neck and legs. Once freed, it rests and dries off. The remains of the yolk, now inside its body, will sustain it for at least a few hours after hatching.

The parent bird disposes of eggshells, either eating them (for the calcium) or carrying them off and dropping them some distance from the nest.

ALTRICIAL AND PRECOCIAL CHICKS

Birds differ significantly in how much care they give their chicks in the days, weeks, or months between hatching and independence, depending on whether the chicks are altricial or precocial.

As soon as its down has dried, a fully precocial chick is ready to leave its nest and will never need to return. It can regulate its own body temperature quite well, run fast and feed itself. A fully altricial chick, though, cannot thermoregulate and must be brooded under its parent's body for several days at least – it must also be fed and cared for because it is unable to leave the nest. A quail's precocial chicks can run after her within hours of hatching and find their own food – they can fly at just 11 days old and leave their mother after three weeks. However, an altricial baby Wandering Albatross (*Diomedea exulans*) remains on or near its nest for about nine months, usually alone as its parents are out at sea, gathering food for it.

Eggs in a clutch may hatch together, or at intervals of a day or more. This depends on whether incubation begins with the arrival of the first egg, or when the clutch is complete. Precocial birds like ducks and geese all hatch together, as the whole family must leave the nest quickly and begin to forage. Many small altricial birds also time their incubation so that all chicks hatch simultaneously. This minimises the total time that they are in the nest (and highly vulnerable), and also means that they will leave it (fledge) at about the same time. They often find separate hiding places soon after

ⓥ Precocial Mallard ducklings enter the water in their first hours of life, already able to swim, dive and catch their own food.

(ⴰ) Adult birds may need to brood their chicks – using their own body heat to provide warmth.

(ⴰ) When chicks hatch asynchronously, the size difference between youngest and oldest can be dramatic.

fledging, and the parents feed them individually until they can feed themselves.

In large birds of prey, which produce semi-altricial chicks (downy, but weak and nest-bound), incubation usually starts with the first egg, resulting in a brood of differing ages. The oldest chick outcompetes the rest for food and so will survive even in lean years, while its younger siblings starve. Total time in the nest is prolonged, but adult raptors can drive away most nest predators. Female Black Eagles (*Ictinaetus malaiensis*) lay two eggs, but the older chick invariably kills the younger one, while the parents look on. The second chick is just an insurance policy, should the first one die in the egg or while very young.

LEARNING SELF-DEFENCE

Young birds are vulnerable to all kinds of predators. Altricial chicks initially respond by calling and gaping when anything touches the nest, but quickly learn to lie down silently if the visitor is not a parent. Precocial chicks stay near their parent when danger threatens. Ducklings dive to escape predators.

Young birds may also have to defend themselves from each other. In most Osprey nests, three chicks hatch and often all three survive, but the youngest can suffer constant violent bullying from the oldest and middle chicks. This treatment may help prepare it for the rigors of life in the wild.

Coot (*Fulica atra*) parents test out their chicks' resilience in the most brutal manner. They often produce large broods, but after a week or two the number is usually down to just two or three, and not necessarily because of predation. The adults will regularly grab and shake their young chicks, and while stronger chicks are quickly released, the weaker chicks will be shaken to death. This 'brood reduction' improves the survival chances for the strongest in the brood.

HOW CHICKS GROW

Embryos grow very rapidly from a single cell to a fully formed, egg-filling chick over as little as 12 days, and then they face the mighty challenge of breaking their way free. From this point, they are no longer passive and helpless.

Altricial chicks are physically very limited at first, although able to demand their parents' care and attention. Though blind, weak and featherless, altricial chicks can raise their heads, open their mouths and make sounds to indicate that they need to be fed. They react to sounds by doing this, as any nearby sound likely heralds the arrival of a parent, and the rest of the time they huddle quietly in the nest to avoid attracting predators. The adults feed them a high-protein diet (even most seed-eating species feed their young on insects) to encourage rapid growth of bones, muscles and feathers. In small songbirds, the weight doubles in the first three to six days of life. The growth rate slows down after the first eight days or so, but the young bird has usually reached adult weight within a couple of weeks.

Among seabirds such as shearwaters, the chick's body weight often overtakes that of its parents. A young Short-tailed Shearwater (*Ardenna tenuirostris*) may be nearly twice as heavy as a typical adult by fledging age, thanks to a high-fat diet of oily fishes. These fat stores enable it to survive when the parents abandon the nest site, and the young bird takes to the seas alone some days later.

Growing feathers

Young birds have only soft, down feathers – these are sparse on altricial chicks but completely cover the bodies of precocial chicks. The contour feathers begin as small bumps in the skin, and these grow into what resemble soft, fleshy pins. Within the pin feather, a blood supply develops, and the barbs of the feather begin to form, while the

ⓛ The down of wader chicks is exquisitely camouflaged so they can rest in relative safety.

ⓥ Altricial chicks can do little more than crane up their necks and beg for food.

indented follicle develops at its base. The barbs eventually break out of the tip of the pin, and the skin that encases them dries, dies, and flakes away as the feather continues to grow. The flight feathers are the first to appear, as they take longer to develop than smaller contour feathers.

Organ development

The bird's internal organs are mostly well-formed by the time it hatches, but they need to increase in size as the bird grows. Many organs' growth keeps step with the body's growth. The intestines' growth rate is governed by food intake, so may be delayed if there are food shortages. The pancreas is one of the fastest-growing organs, while the spleen is one of the slowest, reflecting the relative importance of successful food metabolism versus immune system efficacy in young chicks.

⌃ Some fledgling altricial chicks hop out of their nests before they can fly, relying on their legs to escape from any danger they encounter.

⌄ Not all chicks grow quickly. A baby Black-browed Albatross (*Thalassarche melanophris*) takes about 130 days to grow to fledging age (and another 10 years to reach breeding age).

SPECIALISED CHICK ANATOMY

During the phase of life when they depend on their parents, chicks have particular needs, and corresponding adaptations. Some are anatomical, some behavioural, and most are a combination of both.

Whether precocial or altricial, chicks are equipped with an egg tooth, a hard, sharp growth on the bill tip that is used to break through the eggshell. It is lost soon after hatching.

Before they have hatched, chicks may communicate vocally with their parents, and each other. There is evidence that chicks of precocial species can hear their siblings in the shells and use these cues to time their hatching to be as synchronous as possible. Nearly all young birds have unique vocalisations – in precocial birds, these are to maintain contact with the family, while altricial chicks call to draw their parents' attention to a stomach that needs filling.

Begging calls are loud, and as it begs, a chick cranes its neck upwards and gapes. Each chick in the brood wants to be the one that is fed, and the parent will place the food delivery in the mouth of the most insistent individual. Many songbirds have bright yellow sides to the mouth (gape flanges), which only disappear when the bird has fledged and is feeding itself. In some species (most notably certain Australian finches, such as the Gouldian Finch, *Erythrura gouldiae*) there are prominent swellings and markings inside the mouth to further draw the adult's attention. Chicks of the larger gulls beg by pecking at the red spot on the parent's bill – this stimulates the adult to regurgitate food. This is instinctual – a red spot painted on a stick triggers the same behaviour.

ⓥ A hatchling's bill is soft and slightly rubbery but quickly becomes firmer as the chick matures.

180

(A) Gape markings, as seen in some Australian finch species, help guide a parent to a hungry mouth.

(V) Juvenile plumage may help reduce adult aggression, though not always effectively, as in the case of these American Coots (*Fulica americana*).

CHICK PLUMAGE

Precocial chicks' downy plumage typically has camouflaged markings and colours, and the same is true of the first juvenile plumage of young altricial birds. Until they reach breeding age, they have no need of any of the markings and colours used as signals in courtship and territorial display. The juvenile Robin is entirely brown and spotted. Without the red breast of the adult, it is protected from attack by its parents, as well as being camouflaged. Owls, which often leave their nests while still small and well before they can fly, grow a unique intermediate (mesoptile) plumage of soft feathers that works better than down to keep them warm in the open while their true juvenile plumage is still growing.

Most birds attain adult-like plumage (or almost adult-like) at their first moult, but in some species it takes several years, and

moults, for full adult plumage to be attained. This occurs in birds that take more than a year to reach breeding age. With each moult they become successively more adult-like, but their remaining signs of immaturity may still spare them from serious aggression from full adults. This is important at breeding colonies, which young birds often attend to seek a mate and to observe and learn from adult breeding behaviour.

FEATHERS
AND SKIN

Birds are the only modern animals to possess feathers. These little miracles of biological engineering provide their owners with the means of flight, an insulated covering that can withstand the heaviest weather and a canvas for displaying some of the most beautiful colours and patterns that exist in nature.

> ⊙ The male Mandarin Duck dazzles potential partners with his brilliantly coloured and extravagantly ornamented plumage.

FEATHER TRACTS

Feathers of different types are arranged on a bird's body in a particular way. This arrangement is quite consistent across different families, even though the feathers themselves vary in size, colour and texture.

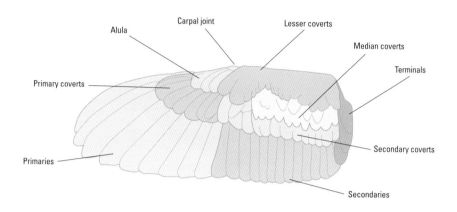

Different groups or tracts of feathers are separated from one another by areas of bare skin, or apteria. The feather tracts, or pterylae, are arranged such that when the feathers are sleeked down they form a continuous 'shell' that smoothly covers the bird's whole body, but if the feathers are raised (for example, when the bird is sunbathing and allowing air to reach its skin) it is possible to see the dense tracts and the apteria between. Feather tracts are also apparent on a baby altricial bird, growing in rows out of the bare skin.

The feather tracts on the wing are most obvious, as the feathers that form them change in size, being smallest close to the body. The tracts on the body are named for the anatomical region they cover. Those covering the head form the capital tract, sometimes divided into the frontal, coronal and occipital regions on top (going from front to back) and the malar region on the cheeks and chin. The spinal or dorsal tract extends from the neck to the base of the tail, running narrowly down the centre of the back and spreading out in front of the hips. The humeral or scapulohumeral tracts run along the sides of the back and inner part of the wing. The femoral tract covers the upper, inner part of the thighs, and the crural tract, the lower feathered part of the leg. On the underside, the ventral tracts run on either side of the breast down the flanks to the undertail. The largest apteria are in the centres of the back and belly.

The wing feathers can be broadly divided into the remiges or flight feathers, which grow from the trailing edge of the wing in a single layer, and the wing coverts, smaller feathers that cover the inner parts of the wing

(∧) Birds all have the same basic feather tracts, and the same general types of markings (such as wing-bars and eye-stripes) are often apparent even in unrelated species. Clockwise from top left: Reed Bunting; Two-barred Crossbill (*Loxia leucoptera*); Willow Flycatcher (*Empidonax traillii*); Sedge Warbler (*Acrocephalus schoenobaenus*); Northern Bobwhite (*Colinus virginianus*).

on both the upper and underside. The tail is structured in a similar way, with the rectrices or tail feathers growing in one layer from the skin at the very tip of the body, and the tail coverts growing on the upper and underside at the base of the rectrices.

DIRECTION AND MARKINGS

Moving from head to tail, the feathers grow pointing backwards, to streamline the bird as it moves forwards, with the front feathers overlapping those behind. As a general rule, the feathers become larger towards the rear of the bird, with the smallest found at the front of the head. Many bird species have distinct markings along the tracts, producing pale wing-bars, dark eye-stripes or rows of spots on the belly, for example. These same general patterns can be found across many families.

FEATHER TYPES

Larger birds may have 20,000 or more individual feathers on their bodies, while small birds can have as few as 1,500. Feathers come in several distinct types with different functions.

(T) The long flight and tail feathers provide lift and control during flight.

The feather is a small miracle of natural engineering. Tracing back its evolutionary pathway takes us to the scales that cover reptiles' bodies, and indeed scales are still evident on modern birds' legs and feet. Feathers are famously lightweight, but also offer very effective insulation, so they are a perfect body covering for an endothermic, flying animal. The feathers on a bird's body, which form its outer surface, are called contour feathers and are adapted (by their shape, and by their arrangement in feather tracts) to provide continuous coverage. Their outer parts are smooth and overlap snugly, resisting wind and water, while close to the skin they are soft and fluffy to trap air, which the skin can warm up. They also provide a canvas for colour and pattern, whether this is intricate camouflage or a riot of brilliant iridescent shades.

As well as the downy fluff at the bases of contour feathers, there are also tiny, entirely downy feathers, which contribute to warming the body. Some birds also have a specialised type of down feathering called 'powder down'. These feathers, which tend to grow close to the belly, crumble to powder, which the bird can then apply to the rest of its plumage when preening. The function of the powder appears to be cleanliness – for example, herons use it to remove fish slime from their feathers – and perhaps also additional waterproofing. Among the contour feathers there are also thread-like filoplumes, which sense when larger feathers are disordered and need to be preened; touch-sensitive filoplumes also grow close to the mouth and eyes. Some insect-catching birds have small, stiff, unbarbed feathers – rictal bristles – lining the edges of their mouths, to help trap prey.

(>) A pair of Greater Racket-tailed Drongo (*Dicrurus paradiseus*). Their elaborate outer tail feathers resemble streamers and lack webbing other than on the tips.

⊘ The red chest marking of bleeding-heart doves (*Gallicolumba luzonica*), which they show off in courtship displays, looks strikingly like a wound.

The long flight feathers of the wings and tail (the remiges and rectrices respectively) are vital for flight and have a distinctive structure. Their shafts are thick and placed towards the feather's leading edge rather than centrally, so that the feather forms an airfoil shape to generate lift. They are extremely strong for their weight but also flexible near their tips. Some birds have emarginated outer primary feathers – these show an abrupt, stepped narrowing near their tips, so that when the wing is fully spread the tips are separated, like fingers, and each acts as an individual miniature airfoil.

DECORATIVE FEATURES

Some feathers show an unusual shape with no obvious practical function, such as the curly tail feathers of a male Mallard, the oversized 'sail' wing feather of a Mandarin, or the flag-shaped outer tail feathers of the Marvellous Spatuletail Hummingbird (*Loddigesia miribalis*). Ornamental feathers like these are almost invariably involved in courtship and territorial display, to be exhibited to rivals and potential mates.

FEATHER STRUCTURE

Feathers have a stiff central shaft that supports the softer parts on either side. The strands along the shaft are fixed strongly to each other through their ingenious structure.

Feathers are made of a protein called keratin, its strands twisted and linked by chemical bonds to form an extremely strong but still very light material. Each feather grows out of a follicle in the skin, and these follicles are connected to one another by skin muscles, which enable the bird to raise or flatten particular feather tracts.

A typical contour feather has several distinct parts. The individual strands that form the main structure of a feather are called barbs. The shaft of the feather has two parts. Its barbless 'stalk', or quill, at the base is the calamus, becoming the rachis from the point where the first barbs are present.

The barbs closest to the calamus are extremely soft and separated from one another, forming the feather's downy or plumulaceous part. The outer part, formed of barbs that are attached to each other to create a smooth, continuous surface, is the vane.

When you handle a feather, you can see the individual barbs in the vane and observe that, while the vane is flexible, the barbs remain attached to one another when you move them gently. You can, however, pull them apart if you apply more force. The barbs are held together in a similar way to Velcro. Each barb is covered in smaller side branches called barbules. The barbules on the upper side of a barb bear tiny hooks, while those on

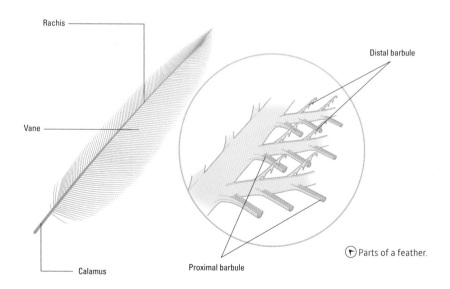

Rachis

Vane

Calamus

Distal barbule

Proximal barbule

Ⓣ Parts of a feather.

(A) Magpies have long tail feathers naturally, but it is not unusual to see individuals that have lost their tails, probably in a battle with a predator. Replacement feathers will grow back in a matter of weeks.

the lower side have tiny grooves. The hooks and grooves 'fasten' together very easily and stay fastened, so each barb can fix to the one above it and the one below. When a bird preens, it refastens barbs that have pulled apart from their neighbors. In the case of most contour feathers, the very tips of the barbs do not fasten together.

LOSING A FEATHER

During growth, a feather has a blood supply in its rachis and if a feather is pulled out, it will cause pain and bleeding. A fully grown feather is not a living structure, though it is held in place in the skin by its calamus until it is either moulted naturally or pulled out by accident. Mature feathers are not so firmly attached that they will not come out of the skin if tugged hard and this can indeed be life-saving – the occasional tailless bird you see probably left its tail feathers in the mouth of a predator.

New feathers will grow immediately to replace any feathers that have been pulled out, provided the follicle has not been damaged. Birds can therefore recover quite quickly from feather loss – even those unfortunate individuals that are left almost bald from a bad outbreak of feather mites can look as good as new within a few weeks.

WORKING WITH WEATHER

Sensitivity to unusual weather patterns and the ability to cope with extreme weather is essential for a bird's survival.

Birds can sense even very slight changes in air pressure, which herald imminent weather change. Falling pressure is an indication of an approaching weather front, which may mean a storm is on the way. Golden-winged Warblers (*Vermivora chrysoptera*) have been shown to sense approaching storms two days in advance, and move out of their path, sometimes going as far as 930 miles (1,500km) to find a safe location. How they sense the change is not yet known but could involve their detecting pressure shifts in the inner ear or within the body's air-sac system, or by hearing distant low-frequency sounds from weather events, which may carry thousands of miles.

Most land birds will still take shelter from heavy rain, but their feather structure and the way the plumage is arranged offer good natural waterproofing. Juvenile birds have less water-resistant feathers and are more at risk of getting their skin wet and becoming too cold. The plumage keeps out the cold as well as water. You will notice how much bigger and fatter your garden birds can look on a cold winter's day, when the feathers are fluffed up to trap and warm up air against the skin.

ⓥ Bee-eaters may struggle to find prey on cool, overcast days when few insects are on the wing.

⊗ Large soaring birds like storks use thermals to gain height without expending too much energy.

⊙ Lacking natural waterproofing, Cormorants and Shags dry their wings after diving by holding them spread in the breeze.

In hot weather, birds pant to cool down, but also sometimes sunbathe, as the sun's heat may discourage parasites. A 'sunning' bird sits with its body plumage fluffed up to allow the sun to warm its skin, and fans out the wings and tail. It simultaneously cools itself by panting with the bill wide open. Cormorants, which lack the natural water-proofing of most diving birds, use the sun and breeze to dry their plumage and raise their body temperature after a session in the water – they sit on exposed perches, facing into the breeze, with their wings held open.

Windy weather can put paid to successful hunting for some birds of prey, but kestrels and barn owls need some breeze to hover efficiently, as it gives them extra lift in this energetically expensive flight style. Soaring birds depend on thermals (rising, circling bodies of warm air) to give them lift – with sufficient thermal activity, they can climb to great heights with barely a single wingbeat. Thermals are vital for opportunist birds like vultures that need to scan a large area of terrain for food, from a great height. These rising, circling bodies of warm air are also key for large migratory birds, many of which cannot make even short sea crossings without first gaining height by riding a thermal along the coastal edge.

Weather and migration

Shifts in weather have a significant effect on migratory behaviour. Winds blowing the wrong way can delay migrants for days or weeks, and bad weather will ground them when they are en route. Even seabirds, well used to rough weather, may be forced to head into bays or even upstream to escape violent sea storms. When migration has to pause, this stimulates the bird to feed rapidly, to restock its depleted fat stores.

BARE PARTS AND SKIN

Most birds' bodies are entirely feathered, except for the feet, lower legs, bill and around the eyes. Bare skin between feather tracts on the body is usually hidden by adjacent feathers.

A bird's skin and other outer coverings (its integument) contain keratin protein. The skin is loose-fitting and flexible in the feather-covered areas. Like mammalian skin, avian skin has an outer layer (epidermis) of dead cells, which wear away but are constantly replaced by new cells that form in the lower layer (dermis). The dermis is rich in blood vessels and sensory nerve endings, and also holds fat stores. The feather follicles and the skin muscles that move the feathers are also present in this layer.

All of the new skin cells gain keratin (become keratinised) as they move towards the epidermis, and also produce sebum, a fatty substance that helps keep the skin soft and pliant. Birds' skin has a simpler structure than mammals' skin, with no sweat glands and no separate sebum-producing (sebaceous) glands.

On the legs and feet, scales of various sizes are formed from plates of particularly keratin-rich epidermis. The spaces between them allow for movement, particularly on the toes. A few birds have entirely feathered legs – this includes many birds of prey, which sport feathery 'leggings'. Others, including many owls and also Ptarmigans and some swifts, even have fully feathered feet, with only the claws exposed. Feathered legs and feet are most common in birds native to very cold climates but may have other functions too. For example, the Ptarmigan (*Lagopus muta*), a ground-feeding bird native to Canada, northern Europe and Eurasia, has feathery feet, which give it a larger and more stable foot area when it walks on snow, in the same way that snowshoes do for us.

Many aquatic birds have webbed feet, formed by extra skin stretching between the front three toes. Webbing may be limited, 'scooping' inwards between the toes, or extend in a straight line from toe tip to toe tip. In some birds, such as pelicans, the hind toe is connected to the rest with webbing too, while in others, such as phalaropes, there are fleshy separate lobes on each toe, rather than toe-to-toe webbing.

ⓐ Webbing between the toes gives a Mallard a pair of large swimming paddles for feet.

ⓐ Moorhens and other rails and crakes have long, unwebbed but fleshy toes that are good for swimming but also for climbing in vegetation.

ⓐ Terns, which rarely swim, have only shallow stretches of webbing between the toes.

ⓐ Coots' toes are not webbed, but have thick fleshy lobes to make them more effective as swimming aids.

FACIAL MODIFICATIONS

In the exposed areas, the integument becomes thicker, tougher and less flexible, sometimes with small warty growths or tubercles. The bones of the bill are covered with firm or hard, thick integument, as are various bony facial modifications, such as the casques of hornbills and cassowaries. Some facial modifications are much softer, though; these include fleshy eye-rings wattles, and the combs of chickens. Bare skin on the face and neck is often used for signalling, such as the colour-changing bare necks of male Turkeys (*Meleagris gallopavo*) or the inflatable red throat pouch of male frigatebirds. The Greater Sage-grouse's bare yellow throat sacs are only revealed in courtship displays.

ⓐ Phalaropes swim much more than other shorebirds, and are the only ones to have swimming lobes on their toes.

MOULT

Feathers are subject to many stresses and need to be replaced regularly. Most birds shed and replace their entire plumage annually, while others undergo more frequent full or partial moults.

Most birds have a set breeding season, at the time of year when conditions are most suitable (in terms of weather, temperature and food availability) for raising a family. For species living in temperate regions, breeding takes place in spring and summer in nearly all cases. The annual moult occurs after breeding, and over a period of a couple of months, if not longer, the bird will have completed its moult and have full, fresh plumage before winter begins. The pattern is more varied and more governed by habitat types in tropical areas, but it is near-universal for adult birds to begin moult around the same time they finish breeding. Juvenile birds usually begin their first moult (the post-juvenile moult) not long after attaining independence.

Partial moults shortly before the breeding season occur in some species, as they transition from a more camouflaged non-breeding plumage to brighter breeding colours. The 'hooded' gull species, such as Bonaparte's Gull (*Chroicocephalus philadelphia*) and Black-headed Gulls (*Chroicocephalus ridibundus*), develop their hoods as a signal of readiness to breed. Changes in day length trigger the body to initiate the process of moult and new feather growth.

Shedding and replacing flight feathers places the bird at risk, as it impairs flying ability. This is overcome, in most species, by the moult occurring very gradually, so that only one or two primary feathers and secondary feathers are missing or not fully

⊙ A male Redstart is brightest in spring, although the plumage is quite worn.

ⓥ Unlike most birds, ducks shed and regrow their flight feathers en masse, rather than one or two at a time.

Multiple loss of flight feathers including bright plumage

Pin feathers begin to develop

Fine feather shafts begin to harden

Fully formed flight feathers, bright plumage returns

grown at any given time. Typically, the innermost primary and outermost secondary are shed first, and then the sequence progresses outwards towards the wing tip with the primaries, and inwards towards the body with the secondaries (see page 184). The remiges or tail feathers are usually shed and replaced from the outside inwards. Some birds, notably many duck species, moult their flight feathers more quickly and as a result have a brief period of flightlessness – these species may migrate to a particularly safe location prior to moulting. Males may also replace their body plumage at this time to a drabber, more camouflaged 'eclipse' plumage, to protect them from predators while they cannot fly.

FROM DRAB TO DAZZLING

Feathers naturally wear away or abrade from the tips inwards over time. This causes most birds to become shabbier in appearance, but it also causes the revelation of colourful breeding plumage in certain species. Birds such as the Common Redstart (*Phoenicurus phoenicurus*) and the Blue Grosbeak (*Passerina caerulea*) have thick pale fringes to their feathers when the plumage is fresh, giving them a generally drab appearance. Over winter, these fringes wear away gradually, progressively revealing the more intense colours on the inner parts of the feathers.

FEATHER CARE

In between moults, a bird's feathers must be kept in the best possible condition. Any shortcomings in their ability to keep the bird warm and dry, and able to fly strongly, could be life-threatening, and, when it comes to attracting a mate, feather quality is hugely important.

Feathers can be affected by many things. The sun bleaches away their darker pigments, feather-chewing lice consume them, unpleasant substances can stain them and affect their waterproofing ability, and they can be caught on thorns and damaged. A blast from a shotgun can rip the sturdiest flight feather in half. However, only if they are pulled out completely, or fall out of their own accord when the bird moults, will damaged feathers be replaced.

Birds therefore devote a lot of time to keeping their feathers clean, dry, parasite-free and properly aligned. When a bird preens, it works its bill systematically through its entire plumage, adopting a series of increasingly contorted poses to access the trickier parts. The long flight and tail feathers are drawn all the way through the bill, from base to tip.

The preening process entails removing any debris or unwelcome invertebrate visitors caught in the feathers, discarding any loose feathers that are caught on their neighbors, 'rezipping' the barbs that have become separated, and aligning any feathers that have become overlapped the wrong way round. Most birds are also applying preen oil to the feathers during this process, which they extract from the uropygial gland on top of the tail base. This oil helps keep the feathers supple and water-resistant. It may also discourage parasites and in some cases imparts a scent that masks the bird's other natural odours, to fool predators. Head

ⓥ An Eastern Bluebird using a garden birdbath. Offering fresh water for drinking and bathing will help attract birds to even the smallest garden.

plumage is difficult to preen with the bill so birds use their feet, and bonded pairs of birds often preen one another. Herons, nightjars and a few other birds have a comb-like serrated edge to one of their claws, which they use when they preen.

(∧) The uropygial gland at the tail-base secretes preen oil, which conditions and waterproofs the feathers.

(∨) Where water is scarce, a dust bath may do just as well.

AVIAN BATHING

Birds also bathe – in water, but also in sand or dust – to remove dirt from their plumage. Arid-country birds such as sandgrouse and larks are avid dust-bathers. Sunbathing is thought to encourage parasites to move to the outer parts of the feathers where they can be more easily removed, and 'anting' – where birds recline on an ants' nest and let the insects crawl over their bodies – is also thought to be an anti-parasite tactic, as the ants squirt stinging formic acid on the feathers. There have been observations of crows 'smoke-bathing', presumably for similar reasons – standing on a smoking chimney and allowing the smoke to pass through their fluffed-out plumage.

PIGMENTATION, PATTERN AND COLOURS

Colours and patterns, whether bright or subtle, simple or intricate, can help a bird both fit in and stand out within its environment, depending on its needs at the time.

⊙ Living in cold northern forests the Ural Owl's (*Strix uralensis*) pale plumage helps camouflage it during the long months of snowy weather.

COLOURS IN THE BIRD WORLD

Few other groups of animals are as colourful as the birds. Feathers can exhibit every shade imaginable, in velvety matte tones or the purest and most brilliant iridescent colours.

The colours of birds and other animals are created in two main ways. First, there is pigment – molecules that are incorporated in the structure of the skin or feathers and impart colour. Then there is structural colour. This is illusory and is produced by the way that light is bounced back from a surface. Its appearance changes with the bird's movements, and with the viewing angle.

A bird's colours are closely linked to its habitat type. The majority of birds need to be inconspicuous, at least for some of the time. Prey species need to ensure they do not catch the eye of predators, and predators need to stay hidden from prey. For this reason, browns and greens are the most common colours we see in land birds, with brown shades most usual in open-country birds and greens predominating in birds that forage in plant foliage. Seabirds usually have

ⓐ The female Pheasant's (*Phasianus colchicus*) dull colours camouflage her when she is incubating her eggs – a task the colourful male does not share.

plumage that combines white, grey and black – this helps conceal them when seen against rippling water.

The most stunningly coloured birds are found in the world's tropical rainforests. Parrots, manakins, tanagers, pittas, hummingbirds, birds-of-paradise and sunbirds show exceptionally vivid colours, across the whole spectrum. In dense, lush vegetation, hiding from view is easy, and it's being seen that is difficult. Bright colours make it simpler for tropical birds to find each other when they want to, and having plumage of iridescent colours means the bird can control its own brightness, as its colours only dazzle when struck by a ray of light.

Ⓐ Most birds that live and forage in tree canopies are predominantly greenish.

Ⓐ Desert species like the Cream-coloured Courser (*Cursorius cursor*) have sandy-toned plumage.

Ⓐ Gulls and most other seabirds are mainly monochrome.

COLOUR AND EVOLUTION

Male birds of open habitats often show what might be considered unwisely bright colouration, making them much more conspicuous than their mates. Here, it is female choice that has driven evolution down a colourful path. In species such as pheasants and ducks, the females alone are responsible for incubation and rearing the chicks, and males are only needed to provide sperm. This setup drives a promiscuous mating system whereby males compete to attract females' attention, and what the females are looking for is not a reliable partner and provider but merely the bearer of strong genes. Males in excellent physical condition will be most popular. Possessing and energetically showing off superbly maintained, colourful plumage demonstrates their fitness, and their ability to evade predators (considerably more difficult without camouflage than with).

Colours are at least partially under genetic control, and genetic mutations can produce aberrant colouration. This may reduce the bird's survival chances but could also unlock new possibilities for living and surviving, and eventually lead to a new evolutionary pathway. Over time, species with a wide geographic distribution display slight differences in colouration in different areas to suit their environment – for example, Peregrine Falcons (*Falco perigrinus*) living in the high tundra are paler than their cousins living in forested southern Asia.

SEXUAL DIMORPHISM

As birds have no external genitalia, telling the sexes apart can be difficult for us (and occasionally even for them). Many birds show no visible sexual dimorphism, and it is only by behavioural traits that an observer can tell which is male and which is female.

In many other species, however, there are marked differences between the sexes in external appearance, and there are sometimes also internal differences, beyond their reproductive anatomy. Among the raptors and owls, females are usually clearly larger than males. This distinction is most marked in those species that hunt other birds – a female Sharp-shinned Hawk (*Accipiter striatus*) may be twice as heavy as a male. This enables 'niche separation' – the female hunts larger birds than the male does, so between them they can prey on a wider range of species. More subtle dietary niche separation between the sexes appears to exist in some other bird groups. For example, in nectar-feeding sunbirds, males and females differ in their efficiency at digesting different kinds of sugars, leading to different feeding choices.

In most birds the male is larger, particularly among those species in which males display and fight communally (lekking) to demonstrate their fitness to watching females. The female Great Bustard (*Otis tarda*) weighs about 4kg (9lb) on average, while males regularly top 10kg (22lb). In such species, males are usually also more colourful with more ornamentation, while the female is drab, as she incubates the eggs and cares for the chicks alone so needs camouflage. The roles are reversed in a few species, such as the phalaropes, in which females are colourful and display to attract males, which are drab and handle incubation and chick-rearing alone.

⌄ In Ospreys, like most raptors, the male is smaller than the female, as it is her role to guard the nest from danger while he hunts for the family.

⊘ A male Sparrowhawk (*Accipiter nisus*) weighs less than 160g (6oz) and mainly hunts sparrow-sized birds.

Clever adaptations

Male songbirds sing more than females do, and their brains have adapted accordingly: male brains have enlarged voice control regions, whereas in female brains the voice perception regions are well developed. The anatomy of the actual vocal system may also be different, for example in ducks, males have larger syrinxes than females, and in certain birds-of-paradise the males possess an elongated and coiled trachea for producing loud, resonant calls.

The uropygial gland (see page 107), present in most birds and the source of preen oil, produces secretions that have a sex-specific odour. Birds of both sexes can discriminate between males and females on the basis of their uropygial scent alone, even in species of songbirds, which are not thought to have a well-developed olfactory sense.

During the incubation period, birds develop a brood patch, an area of skin on the belly that sheds its feathers and becomes swollen, with a denser blood supply. The brood patch transfers more of the bird's body heat to its eggs as it incubates. Because in most species it is the female that carries out most if not all of the incubation, the brood patch is usually a female trait.

In some species, competition for resources outside of the breeding season may lead to one sex having a different distribution to the other. In Eurasian species like the Pochard (*Aythya ferina*) and the

⊙ The female Sparrowhawk weighs about 260g (9oz), and her prey consists mainly of dove-sized birds.

Chaffinch (*Fringilla coelebs*), the slightly smaller females are forced to move further south than males in order to survive the winter, resulting in single-sex flocks in some areas.

PIGMENTATION TYPES AND FORMATION

Pigments are 'true' colours, deposited as part of the molecular structure of feathers and skin. Some can be formed naturally within the body, while others cannot and come from the diet.

The most common pigment found in birds' feathers is melanin. This pigment comes in two distinct forms – eumelanin and pheomelanin. Some birds have only eumelanin, while many birds have both types. Eumelanin gives black, dark grey and dark brown tones, while pheomelanin is responsible for lighter brown and reddish tones. The colours of the American Robin and the European Robin (both of which are darkish brown birds with red breasts) derive from a combination of eumelanin and pheomelanin.

Melanin molecules are created in melanocyte cells. These cells are found in the bird's skin, and contain specialised organelles called melanosomes, which build, store and transport melanin molecules. The construction depends on the enzyme tyrosinase, which encourages the amino acid tyrosine to react with oxygen. Several further

ⓐ The male Northern Cardinal's stunning red plumage is produced by carotenoid pigment.

stages of chemical reaction take place to produce melanin molecules. These combine into larger granules, which can contain both types of melanin. The granules are passed into the cells of growing feathers. As well as imparting colour, melanin makes the feather stronger and less resistant to wear. This is why, in many birds, the tips of the flight feathers are darker than the rest of the plumage – the most striking example is the black wing tips of otherwise pale-plumaged gulls.

As feathers wear over time, their melanin colours fade, turning blacks to greys. Other pigments are less susceptible to fading.

(∧) Less vivid reds, like the breast of an American Robin, can be produced by pheomelanin.

COLOUR BOOSTERS

Carotenoids are another class of pigment, which give rise to bright yellow, orange and red tones. They are present in birds such as the Northern Cardinal (*Cardinalis cardinalis*) and the Yellow Wagtail (*Motacilla flava*). Carotenoids are created in plants and passed on to the bodies of animals that eat those plants – most birds obtain their carotenoids via plant-eating insects, such as caterpillars. Because carotenoids can only be obtained this way, there can be considerable variation in the brightness of carotenoid-derived

colours within the same species. In the Great Tit (*Parus major*), which has a yellow belly, the brightness of the yellow is an 'honest signal' of fitness – it shows the bird is well-nourished, and the brightest birds have the most breeding success.

Melanins and carotenoids often occur in combination, to dazzling effect. The male Blackburnian Warbler's (*Setophaga fusca*) plumage shows a striking mix of black and yellow patches, with an intensely orange throat. The two pigment types also create shades of green when they are deposited in the same individual feathers, as can be seen in the Greenfinch (*Carduelis chloris*).

A third, less common pigment class, the porphyrins, are synthesised in the body from amino acids. They produce various tones including pink and green, but are notable for shining bright red under ultraviolet light. They are found in turacos and some pheasants, pigeons and owls.

(↱) A combination of black eumelanin and bright yellow carotenoids produces the dazzling colour scheme of the Blackburnian Warbler.

STRUCTURAL COLOURS

Iridescence is caused by the interplay of light on feathers that have a particular structure. Their colour appears to shift, usually through bright greens, blues and violets, depending on the angle of light, but this is illusory – the actual pigmentation is usually the black, grey or brown of melanins.

In school science classes, most of us will have seen how a beam of white light shone into a glass prism will be deviated and split into a rainbow of its component colours. This phenomenon, the 'bending' and scattering of light waves, is called refraction. Iridescent feathers work in a similar way. The keratin in the barbules is arranged in corrugated layers, and these refract incoming light, splitting it into its constituent colours. Some of those colours are absorbed by the barbule's melanin layers, while others are reflected back, primarily those of shorter wavelength (blues, violets and greens), but the ones we see change constantly, depending on our viewing angle.

Another kind of structural colouration is produced by the scattering of light, rather than refraction. These feathers have a different structure, incorporating a 'spongy' keratin layer in the barbules that incorporates tiny air pockets. These air pockets reflect back particular colours of light. As with iridescence, the underlying pigmentation of the feather is melanic, and other light colours are absorbed by the melanin layers. Unlike iridescence, colour produced by scattering is always the same shade, and does not necessarily have a glossy appearance, though it can appear to 'shimmer' brighter in certain light conditions. Many pure blue tones in birds' plumage, for example in the male Indigo Bunting (*Passerina cyanea*), or on the crown of the Blue Tit (*Cyanistes caeruleus*), are produced by scattering.

SPECTACULAR COLOUR

The most spectacularly coloured birds on Earth show pigmentation and structural colour in combination. Among them is the Paradise Tanager (*Tangara chilensis*), with its lime-green face mask, deep-blue throat, sky-blue belly and bright red rump, set off to stunning effect against its jet-black wings and tail. The Many-coloured Rush Tyrant (*Tachuris rubrigastra*), unlike most of its

⊲ The Paradise Tanager's glistening blues and greens are examples of structural colour.

The light-scattering effect produces the deep blue of male Indigo Buntings' plumage.

The Many-coloured Rush Tyrant's colours are formed by melanins, carotenoids and structural factors.

drab-plumaged relatives in the tyrant-flycatcher family, is resplendently yellow below and vivid green above, with iridescent bright blue cheeks and a scarlet undertail. The Kingfisher, one of Europe's brightest birds, owes its vivid blue and deep orange palette to a combination of melanin pigment and both kinds of structural colour.

Structural colours exist elsewhere in the animal kingdom, most notably in certain beetles and butterflies. As with birds, the brightest examples are typically found in tropical, forested areas where a flash of colour stands out in the otherwise rather dark environment.

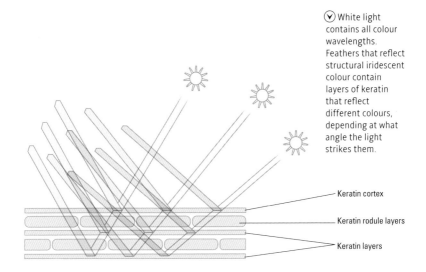

White light contains all colour wavelengths. Feathers that reflect structural iridescent colour contain layers of keratin that reflect different colours, depending at what angle the light strikes them.

Keratin cortex

Keratin rodule layers

Keratin layers

ABNORMAL COLOURS

Odd-coloured birds occasionally occur in the wild, usually because of random genetic mutation. Such birds rarely survive to breeding age to pass on their genes, mainly because they lack natural camouflage.

When a feather that would normally contain melanin or other pigment has none, it will reflect all light wavelengths and so appear to be pure white. Genetic mutations can cause disruption to the creation and deposition of melanin, which may affect the bird's entire plumage, or just parts of it. If a bird that normally has only melanin pigments does not exhibit any melanin at all, including in its eyes and other bare parts, it is white with pink eyes, bill and legs, it is an albino. It is very rare to see albinos in the wild, possibly because a lack of melanin in the eyes has a negative effect on eyesight. But birds with normal eyes and white feathers are more frequent. This is usually known as leucism (partial leucism if only some feathers are affected).

In other cases, feathers may have reduced melanin, producing birds with 'dilute' plumage. Other abnormalities include lacking one of the two melanin types. In the American Robin, lack of pheomelanin results in a greyish bird with a white rather than red breast, while lack of eumelanin makes for a red-breasted but otherwise whitish bird. Excess melanin gives a bird darker plumage than normal (melanism).

The green colour of wild-type Budgerigars (*Melopsittacus undulatus*) is a combination of structural blue and carotenoid-pigment yellow. Budgies that are unable to deposit the yellow pigment in their feathers are blue with white faces, while those that lack melanin (which is required for the light-scattering effect of structural colour to work) will be pure yellow, or lutino. Genetically, lutino is the same as albinism – the bird has a total loss of melanin – but the yellow colour is unaffected because it comes from carotenoids. All of these colours have arisen spontaneously in captive Budgerigars, and then been propagated by breeders.

ⓒ A normally pigmented Vermilion Flycatcher has bright orange and brown plumage.

COLOUR AND NUTRITION

Birds sometimes show temporary pigmentation problems that are not genetic in origin but caused by nutritional deficiency. Flamingos in captivity often fade from pink to white because artificial diets are too low in carotenoids, which the birds derive from eating shrimp. You may occasionally see wild juvenile Carrion Crows (*Corvus corone*) in urban England with white bands in the

Ⓐ Vermilion Flycatchers (*Pyrocephalus rubinus*) with melanism are common in Lima, Peru – their dull plumage gives them a survival advantage in this heavily polluted city.

Ⓐ One of the most common aberrations in birds' plumage is partial leucism, where some feathers lack their normal pigmentation and are white.

centres of their flight feathers, due to a substandard diet while in the nest, but they will replace these with normal feathers at their next moult provided they eat well in the meantime.

Bird-keepers that breed Canaries (*Serinus canaria domestica*) have introduced new colours to some breeds by giving the birds a diet rich in carotenoids. Adding highly pigmented foods, such as carrots and beetroot, to the bird's normal seed is effective. These 'red-factor' Canaries can attain extremely deep, rich shades of orange or reddish, but the colour cannot be passed on to the birds' offspring. If the colour-feeding stops, the birds soon revert to normal colouration.

CAMOUFLAGE AND ILLUSION

While bright colour has its place, many birds prioritise invisibility and have made camouflage an art form, whether they need to be unseen by predators or prey, on land or at sea.

⊙ Woodcocks are extremely hard to spot as they feed on a forest floor.

Birds that nest or roost in exposed situations often have extraordinarily detailed and intricate camouflaged markings. Among them are owls, nightjars and woodcocks. Their brown, grey and cream colours are formed mainly by melanins. Individual feathers often have dark shafts and lighter and darker crossbars. The breast feathers of owls typically have jagged vertical streaks, which resemble the fissures on tree bark. They can exaggerate the effect by tightening their plumage and standing upright. Owls with pointed 'ear tufts' may resemble the snapped-off stump of a vertical branch when they perch midway along a horizontal branch (see page 84).

Ground-nesting woodland birds are drab and mottled, to give them colour camouflage and to disrupt their outline, making them less obviously 'bird-shaped'. However, those that feed and nest in foliage are more often plain greenish. Species that nest on the ground in more open habitats, such as dunes and deserts, have plainer, paler, sandy-toned plumage. The light brown but dark-banded plumage of the Ringed Plover (*Charadrius hiaticula*) and its relatives camouflages it well on shingle beaches. Nearly all of these birds have darker upper sides and paler undersides. This 'counter-shading' works against the natural shadowing of the underside when the bird is lit from above, making its form less obviously three-dimensional.

Camouflage while on the move is more difficult to achieve, but can be seen with impressive effect in some seabirds. Prions

 Ringed Plovers have bold plumage patterns, but this serves to break up and disguise their outline on a shingle beach.

and many other petrels, as well as young gulls of certain species, show a distinctive zigzagging, dark M-shaped marking across the wings when seen from above. This marking mimics ripples in the sea surface, making the bird hard to spot from above and so less likely to fall prey to larger, predatory seabirds. Most seabirds have white undersides, which makes it harder for their underwater prey to spot them from below as they fly against a bright sky.

Many birds that forage alongside running water have a black-and-white pattern that breaks up their shape against the moving shadows on the water. Some birds adopt distinctive postures for extra camouflage, such as the American (*Botaurus lentiginosus*) and Eurasian Bittern (*Botaurus stellaris*), a reedbed bird which points its bill upwards when alarmed, so that its throat streaks disguise it against the reeds.

THE ART OF DECEIT

Some birds have markings intended to deceive rather than conceal. The Sunbittern (*Eurypyga helias*), a drab wading bird, reveals markings that resemble two huge eyes when it spreads its wings – enough to alarm a would-be predator and allow the Sunbittern to escape. Many small owls have 'false eye' markings on the backs of their heads, to suggest that they are looking in both directions at once and cannot be stalked.

 The Sunbittern's spread wings reveal startling large eye patterns, suggesting a much bigger animal.

ORNAMENTATIONS

Spectacular-looking birds are often those with elaborate 'extras' in their appearance. The huge, many-eyed fantail of the Peacock, the inflatable scarlet balloon-throat of the Magnificent Frigatebird (Fregata magnificens), *even the comb and wattles on a domestic rooster – these are ornamentations with no apparent function beyond the decorative.*

Most of the birds we know well are quite straightforward-looking – sleek and unadorned. This makes perfect sense for an animal that has to move fast and with efficiency. Adornments are potential encumbrances. Yet some birds do have these encumbrances and they probably do have an appreciable impact on survival. A male Long-tailed Widowbird's (*Euplectes progne*) keel-shaped, elongated tail feathers are more than twice as long as his body – a significant extra weight for a small bird to carry around, a breeze-catching hindrance to flight, and an easy target for a predator to grab at.

It is no coincidence that it is usually male birds that have ornamentations, and that they use these in their courtship displays. The Long-tailed Widowbird attracts females by leaping high out of long grass, the tail forming a distinctive shape in the air. Male Turkeys' bare faces turn vivid blue and red or bright white as they strut, fight and display communally for watching females; the colour change is caused by changes in the blood supply to the skin, which in turn affects the way light refracts from its surface.

Other ornamentation includes the large yellow bulge, boldly outlined in black, on the bill-base of the King Eider (*Somateria spectabilis*) duck, the elongated and brilliantly iridescent throat feathers ('beards') of some hummingbirds, and the long, inflatable wattle that hangs down from the breast of male

ⓥ Male King Eiders' ornamental bill 'knobs' are noticeable over long distances.

212

⊙ The male Long-tailed Widowbird forms a striking shape in its leaping display.

⊙ Hormonal excitement turns a male Turkey's facial ornamentations vivid shades of red and blue.

⊙ Both sexes of the Masked Lapwing (*Vanellus miles*) have facial wattles, but the male's are more prominent.

umbrellabirds. The bowerbirds, related to birds-of-paradise, are not themselves ornamented but collect colourful items and use them to decorate arenas or nest-like structures – their bowers.

Female choice – sexual selection – is what drives these characteristics. Any male that can survive despite the burden of extreme ornamentations must be exceptionally fit, and so have strong genes to pass on. The most elaborately ornamented birds are indeed also among the healthiest and most long-lived, with bodily resources to spare. Of course, if this selective pressure modifies male bodies enough to seriously harm their survival chances, the most extreme males will die before they can breed, so natural selection ultimately trumps sexual selection.

Functional ornamentation

Some other apparent ornamentations, found in species without sexual dimorphism, include crests and ear tufts. These, however, are often not ornamental but enhance camouflage by giving the bird a less 'bird-like' silhouette. The enormous casques on hornbills' upper mandibles are another example of an apparent ornamentation with a practical function – they increase resonance of the bird's calls, and also strengthen the bill, allowing for more pressure to be applied by the mandible tips when the bird picks up a fruit or other food item.

BARE PART COLOURS

Colour is not confined to feathers. Some birds have highly coloured bare facial skin, bills, eyes, legs and feet, and even the insides of their mouths can be startlingly vivid.

Many otherwise drab birds have colourful bills, legs, or both. These body parts can usually be concealed fairly easily – for example, a Common Redshank (*Tringa totanus*) on its nest will be covering its scarlet legs with its feathers. Colourful bare parts can, therefore, be a low-risk way of signalling to others of the same species. Among those species with brightly coloured legs and bill, the most frequent colours are red, orange and yellow. Most gulls and terns, even though otherwise monochrome, have colourful bills and/or legs. The colour they show is derived from carotenoids, which (just as with carotenoid feathers) come from their diet. The brightness is therefore a reliable sign of the bird's condition – those with the brightest colours will be those that are most successful at finding nutritious food. They will tend to prevail over their less bright rivals in any conflict over territory, and will be preferred as breeding partners as they are

ⓐ The Long-tailed Tit has striking pink eyelids, changing to deeper red in moments of agitation.

more likely to survive the season and take good care of their young.

The skin around the eye may also be conspicuously colourful. The gannets have strikingly bright blue eye-rings, with black bare skin around them, creating a boldly

ⓓ A Keel-billed Toucan's oversized bill is a fruit-plucking tool but also provides a canvas for vivid colouration.

marked face. Their relatives, the Blue-footed and Red-footed Boobies (*Sula nebouxii* and *Sula sula*), have very brightly coloured legs and feet. All of these birds are noted for engaging in elaborate courtship displays in which they show off their colourful bare parts.

The toucans, gregarious birds of the South American rainforest, are exceptional in the size, colour and patterns they show on their extremely large bills. The Keel-billed Toucan (*Ramphastos sulfuratus*), for example, has a green bill with a red tip, marked with blue on the lower mandible and an orange stripe along the upper mandible. The bird as a whole is not highly coloured, being black with a pale-yellow face and bib. The bill, used to reach fruits on twig tips, is huge but very lightweight, and is key to keeping the bird cool (through dilation of its blood vessels), but there seems to be no particular explanation for its striking colour scheme beyond a visual contact cue for other members of the flock.

FACIAL SIGNALS

Some bare-faced birds change the colour of their facial skin according to their mood, providing a visual signal to others of their species. Displaying male Turkeys' faces change from pink to blue or white when they are very excited and ready to confront a rival. The pink eyelids of Long-tailed Tits (*Aegithalos caudatus*) flush darker reddish when they are alarmed – these small birds travel in family groups and are in constant vocal as well as visual communication.

(⌃) In its solemn courtship dance, the Blue-footed Booby makes sure its mate gets a good look at its colourful feet.

(⌄) Gannets (*Morus bassanus*) show off their blue eye-rings in their face-to-face bill-tapping courtship display.

GLOSSARY OF TERMS

Abrade Wear away

Adaptive radiation When one species diversifies into many over time, usually because new ecological niches are available

Air cell Air-filled space inside an egg

Air sacs System of membranous pouches connected to the lungs

Albumen Egg white, or the protein contained in it

Allantois External circulatory system of an embryo

Alula The 'thumb' of the wing

Amino acids Small molecules that join to form proteins

Amniote Amniotes lay their eggs on land or retain the egg inside the body of the mother

Apteria Region of bare skin between feather tracts

Arteriole Small blood vessel that branches from an artery

Artery Blood vessel leading away from the heart

ATP (Adenosine triphosphate) Molecule broken down in cell mitochondria to release energy

Barb Side branch of a feather

Barbule Side branch of a barb

Bill (beak) Jaws or mandibles, made of bone with a tough keratin sheath

Call Simple sounds produced by birds for communication

Capillaries Tiny, semi-porous blood vessels within tissues

Carotenoid Pigment giving yellow, orange and red tones

Cartilage Tough, inflexible connective tissue

Chalaza Protein layer around egg yolk that twists into supportive strands

Chromosomes Paired strands of DNA in a cell nucleus

Cloaca Opening for copulation and excretion

Cochlea Sound-sensing part of the inner ear

Collagen Protein molecule found in connective tissue

Contour feathers Feathers that cover the main part of the body

Convergent evolution When two unrelated animals have similar anatomy and physiology because of similar lifestyles

Crop First part of the digestive tract, where food is stored and softened

Dermis Skin layer beneath the epidermis

Display Exaggerated, ritualistic movements performed to intimidate rivals and attract mates

Duodenum First section of the small intestine

Eclipse Drab plumage of some male birds, seen during moult

Egg tooth Temporary hard growth on the bill tip of a chick used to crack the eggshell

Enzyme Molecule that helps break down food

Epidermis Outer skin layer

Erythrocyte Red blood cell

Fatty acids Small molecules that join to form fats

Feather tract Discrete grouping of similar-sized feathers

Fermentation Breaking down food through the actions of bacteria

Filoplumes Small hair-like feathers used to sense touch

Furcula Fused collarbones of a bird

Gamete Sex cell (sperm or ova)

Germinal disk Region of an egg where the embryo develops

Gizzard Muscular digestive organ, for food breakdown

Gland Organ that secretes hormones or other substances

Hormone Protein molecule that stimulates particular cell activity

Insectivore Feeding mainly or exclusively on insects and other small invertebrates

Integument Protective outer covering (skin, feathers, bill etc.)

Iris Coloured part of the eye, a sphincter muscle that lets in light

Keel Ridge on a bird's breastbone

Keratin Protein found in feathers and skin

Leucocyte White blood cell

Ligament Cartilage linking bones to other bones

Lymph Fluid found between cells and carried in the lymph system

Lymphocyte Type of white blood cell involved in acquired immune responses

Mandible Lower jaw

Melanin Dark pigment

Mitochondria Organelles in a cell that generate energy

Moult Shedding worn feathers and growing replacements, usually annual

Neuron Nerve cell

Niche Particular environment (including its resources) that can support a particular species

Oestrogen Hormone driving female-typical physiological activity and behaviour

Organelle Discrete structure within a cell

Ovary Egg-producing organ of a female bird

Oviduct Female reproductive tract

Pallium Outer brain layer in birds, involved in higher mental functions

Patagium Thin skin membrane on the forelimb, used for gliding or flight (for example, in bats)

Pectoral muscles Muscles in a bird's chest that drive its wingbeats

Pellet Compacted mass of indigestible food remains, regurgitated after eating

Phagocyte White blood cell that works by engulfing pathogens

Pigment Molecule that gives colour to feathers and other tissues

Plasma Liquid component of blood

Post-juvenile moult Moult of a young bird to replace its first (juvenile) feathers

Preen oil Secreted by the uropygial gland, for feather care

Preening Cleaning and repositioning feathers using the bill

Primaries/primary feathers Outermost, longest rectrices

Proventriculus First part of the bird's stomach

Pygostyle Fused tail vertebrae of a bird

Rachis Main shaft of a feather

Rectrices Long tail feathers

Remiges (flight feathers) Long wing feathers

Retina Membrane at the back of the eyeball, containing light-sensing rod and cone cells

Ringing (banding) Uniquely marking individual birds with numbered leg-rings

Rostrum Top part of the bill, ormed by the upper jaw and nostrils

Scales Hard plates on a bird's legs and toes

Secondaries/secondary feathers Inner, shorter rectrices

Skeletal muscle Muscle tissue found in muscles between bones, allowing voluntary joint movement

Smooth muscle Muscle tissue that lines some internal organs and vessels, allowing involuntary contraction

Song Sounds produced by birds to advertise for a mate and declare territory ownership

Sphincter muscle A ring-shaped muscle that can tighten or open

Structural colour Colour produced by refraction or scattering of light rather than pigment

Syrinx Organ of vocalisation, located at the base of the trachea

Tail coverts Shorter feathers on the basal part of the tail

Tarsometatarsus Fused ankle bones, forming the longest visible part of a bird's leg

Tendon Cartilaginous end of a muscle, where it connects to a bone

Testis (plural testes) Sperm-producing organ of a male bird

Testosterone Hormone driving male-typical physiological activity and behaviour

Tetrapod Vertebrate animal with four limbs

Theropod Two-legged, often feathered dinosaur

Torpor State of much-reduced metabolic activity

Trachea Windpipe, linking lungs to the mouth

Uropygial gland Gland on the tail base that produces preen oil

Vane Main part of a contour or flight feather, where barbs are joined to form a continuous, air-resistant surface

Vein Blood vessel leading back to the heart

Venule Small blood vessel that joins a vein

Webbing Thin skin between the toes

Wing coverts Shorter feathers on the inner part of the wing

INDEX

INDEX

INDEX

PICTURE CREDITS